BETA DECAY for PEDESTRIANS

HARRY J. LIPKIN
The Weizmann Institute of Science
Rehovot, Israel

DOVER PUBLICATIONS, INC.
Mineola, New York

Bibliographical Note

This Dover edition, first published in 2004, is an unabridged republication of the work originally published in 1962 by North-Holland Publishing Company, Amsterdam, and Interscience Publishers, New York.

Library of Congress Cataloging-in-Publication Data

Lipkin, Harry J.
Beta decay for pedestrians / Harry J. Lipkin.
p. cm.
Originally published: Amsterdam : North-Holland Pub. Co. ; New York : Interscience Publishers, 1962.
Includes bibliographical references and index.
ISBN 0-486-43819-8 (pbk.)
1. Beta decay. 2. Radioactivity. I. Title.

QC793.5.B425L57 2004
539.7'523—dc22

2004056236

Manufactured in the United States of America
Dover Publications, Inc., 31 East 2nd Street, Mineola, N.Y. 11501

PREFACE

A bulletin entitled 'Parity for Pedestrians' was circulated by the present author at the Rehovoth Conference in 1957, at the time of clarification of the confused experimental situation following the discovery of parity non-conservation. This bulletin attempted to describe in simple language the essentially simple experimental results and their theoretical implications. The 'pedestrian approach' was developed further in a series of lectures at the International Summer Meeting at Tučepi, Yugoslavia, in 1958; in lectures to the Frauenfelder group at Illinois in 1959; and in lectures at the Weizmann Institute in 1960.

The aim of the approach is to present simply, but without oversimplification, those aspects of beta decay which can be understood without reference to the formal theory, i.e. those relations which follow directly from conservation laws and elementary quantum mechanics. The pedestrian treatment is not intended as a substitute for a complete treatment nor as a watered-down version.

A derivation of results obtainable without the formal theory should make these results more understandable and less mysterious to those who have neither the time nor the inclination to master the details of the theory.

On the other hand, those who know the formal theory should find in the pedestrian treatment a clear distinction between results which depend upon the specific assumptions underlying the formal theory, and those which are independent of these assumptions and follow from simple general principles. At a time when peculiar experimental results may still arise, it is useful to be able to see in a simple way which assumptions or approximations used in the theory are challenged by each result.

The author would like to express his appreciation to M. Mladjenovic and his colleagues for the hospitality of the Yugoslav Summer Meeting, to members of the staff of the University of Illinois, particularly J. D. Jackson for stimulating and critical discussions, and to M. Peshkin, P. Hillman, S. Meshkov and A. Nir for discussions and criticism of the manuscript.

CONTENTS

CHAPTER IV

ANGULAR MOMENTUM IN BETA DECAY

CHAPTER V

ALLOWED BETA DECAY WITH FORMULAS

CHAPTER VI

CONNECTION WITH BETA DECAY THEORY - CONCLUSIONS

CHAPTER 1

INTRODUCTION

1.1. DIRAC MATRICES AND RACAH ALGEBRA

After the discovery that parity was not conserved in beta decay, an extensive series of experiments was undertaken in many laboratories to determine the nature of the beta decay interaction. For a general review of these experiments and a detailed bibliography see KONOPINSKI [1959] and SCHOPPER [1960]. These were mainly measurements of the angular distributions and polarizations of particles emitted in and after beta decay. Many calculations were made giving the results of these experiments as a function of the parameters (coupling constants) characterizing the beta decay interaction. In general these calculations involved complicated manipulations of Dirac matrices and extensive use of the algebraic techniques developed by Racah for coupling angular momenta. On the other hand, it was often possible to obtain considerable insight into the physics of a particular experiment by simple geometrical arguments, usually accompanied by hand waving, drawing pictures, and describing particles as 'spinning *this* way' (or *that* way) and going 'up' (or down). Despite the success of these arguments the general feeling prevailed among those in the field that they were 'good for experimentalists', not quite respectable, and that the right way to treat the problem was with the full paraphernalia of Dirac matrices and Racah algebra.

The purpose of this little book is to show that these simple arguments can be put on a rigorous, respectable basis, and to develop them as much as possible. We shall see how much beta decay physics can be described and understood without using Dirac matrices and Racah algebra. Perhaps the reader will feel some of the surprise and delight felt by the author in first learning how far one can go with this simple approach.

The author bears no prejudice against Dirac matrices and Racah algebra. They are interesting, useful, and certainly necessary for a full description of all the aspects of beta decay. However, considerable physical insight is gained by exploring the extent to which the experiments can be understood without Dirac matrices and Racah algebra, particularly in the allowed transitions.

In doing without Dirac matrices and without the Dirac equation, we are essentially doing completely without what is usually called the theory of beta decay. All results obtained in this way, all the relations obtained between the results of various experiments, are therefore independent of conventional beta decay theory. Our treatment will therefore show what kind of experimental information is required to test beta decay theory, and what kind of experimental information is independent of beta decay theory and really tells us only the momentum and angular momentum are conserved.

In doing without Racah algebra we are not really doing anything profound. It simply happens that in allowed beta decay the angular momenta involved are so small (both the electron and neutrino have total angular momentum one-half) that all the appropriate angular momentum coupling coefficients reduce to trivial cases. It is really a shame to use such a powerful and elegant method for such a simple problem. We shall instead present a rigorous treatment of angular momenta which is simply related to the hand-waving picture-drawing arguments.

In the remaining sections of this chapter, we present a summary of the simple picture-drawing arguments, the conclusions which can be inferred, and an analysis of the underlying physical basis. In the following two chapters we develop these arguments into a rigorous theory of beta decay without Dirac matrices, demonstrating all results which are independent of beta decay theory. These results are complete except that the exact explicit form of the angular distributions is not given. In the next two chapters, the necessary properties of angular momentum are developed and applied to obtain exact quantitative results. Finally the predictions of beta decay theory are discussed in the light of the preceding simple treatment without going into the details of the theory, and the present experimental situation is discussed.

1.2. PROPERTIES OF THE ELECTRON AND NEUTRINO EMITTED IN BETA DECAY

In beta decay a nucleus emits an electron and neutrino and is transformed into another nucleus of the same mass number and having a charge differing by one unit from that of the initial nucleus (or a nucleus can capture an orbital electron and emit a neutrino).

$$\mathrm{N}(Z, A) \rightarrow \mathrm{N}(Z \pm 1, A) + \mathrm{e}^{\mp} + \nu$$
$$\mathrm{N}(Z, A) + \mathrm{e}^{-} \rightarrow \mathrm{N}(Z - 1, A) + \nu\,.$$

In beta decay experiments it is generally some properties of the emitted leptons, the electron and neutrino, that are measured, such as their energy, polarization or angular distribution. For this reason it is of interest to examine the possible properties and states of these leptons.

Both the electron and neutrino have spin one-half. There are therefore two possible orientations for each spin. We can describe these two states in several ways. If the motion of the particle in space is described by a plane wave, we can talk about 'spin up' or 'spin down'. If the particle has a well-defined total angular momentum j, we can say that there are two possible states for each value of j with 'opposite' spin orientations, one in which the orbital angular momentum l is $j + \frac{1}{2}$, and one in which l is $j - \frac{1}{2}$. Since the orbital angular momentum is not a constant of the motion for a relativistic particle, but *parity* is, we can modify this classification by saying that there are two states for the same j, one with *odd* parity and one with *even* parity. In the non-relativistic limit this reduces to the case where the two states have a definite orbital angular momentum. For example, if $j = \frac{1}{2}$, then the orbital angular momentum can be zero or one (s-state or p-state). The states of even and odd parity are both mixtures of $s_{\frac{1}{2}}$ and $p_{\frac{1}{2}}$ with the degree of mixing depending upon the energy. In the extreme relativistic limit they are both equally mixed; in the non-relativistic limit they are separated with the s-state having even parity and the p-state odd.

In beta decay, angular momentum is conserved but it is now known that parity is not conserved. The leptons are emitted in states of indefinite parity. Thus it is not convenient to use parity to distinguish between the two states of the same angular momentum j. As longitudinal polarizations of particles are frequently measured, a more suitable way to specify the spin direction is in terms of the longitudinal polarization, or helicity. This classification is commonly used with circularly polarized light quanta. We thus have 'left-handed' states in which the direction of the spin is opposite to the direction of motion and 'right-handed' states in which the direction of the spin is parallel to the direction of motion.

In allowed beta decay transitions, the electron and neutrino are both emitted in states of total angular momentum $j = \frac{1}{2}$. There are therefore two possible directions for the orientation of the total angular momentum with respect to axes fixed in the laboratory. The projection of j on the fixed axis is usually denoted by m and can take on the values $m = \pm \frac{1}{2}$.

In allowed decay each lepton has four possible states. It can be either left or right-handed, and its angular momentum can be pointing either 'up' ($m = +\frac{1}{2}$) or 'down' ($m = -\frac{1}{2}$). There are thus 4×4 or sixteen possible states for the combined electron-neutrino system. These states are represented graphically in Fig. 1.1.

1.3. PICTORIAL PREDICTIONS OF EXPERIMENTAL RESULTS

Fig. 1.1 shows the sixteen possible electron-neutrino states of allowed beta decay. In each diagram the lines indicate the paths of the electron and neutrino emitted from the nucleus; the arrows indicate the direction of the spin. Each path is labeled e or ν indicating that the particle is an electron or neutrino. The subscripts R and L indicate that the particle is right-handed or left-handed. The diagrams are also labeled by the projections of the electron and neutrino angular momenta on a vertical axis, m_e and m_ν.

Each diagram in Fig. 1.1 indicates a definite direction of emission for each lepton with respect to the vertical axis, and therefore with respect to the direction of emission of the other lepton. This is equivalent to the simple 'hand-waving' arguments like the one that a particle whose angular momentum is directed upward ('it's spinning *this* way') and which is *right*-handed, must be moving *up*. It is a trivial result of the definition of 'right-handed' or 'left-handed' longitudinal polarization (or *helicity*) as the sense of the projection of the spin on the direction of motion. If any two of these three quantities is known, spin

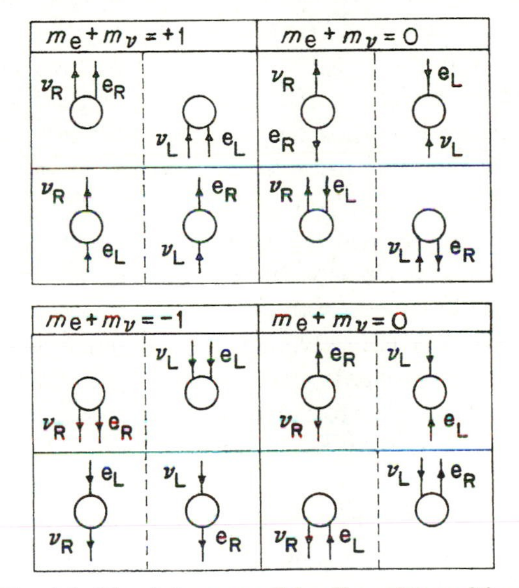

Fig. 1.1. The sixteen possible allowed transitions.

direction, momentum direction, and polarization, the third is uniquely determined by this definition. We can thus express any one of the three quantities in terms of the other two. This trivial result is used extensively in the simple arguments interpreting allowed beta decay experiments.

Let us now examine the three common types of experiments measuring angular distribution in allowed beta decay.

1. *Angular distribution of electrons emitted from polarized nuclei.* This was the first experiment performed which showed that parity was not conserved in beta decay (WU [1957]). We take for simplicity the case which was first measured, that of Co^{60}. The spin of Co^{60} is 5, and it decays by negaton emission to a state of Ni^{60} which has a spin of 4. The cobalt nuclei are polarized by a magnetic field at low temperature and the angular distribution of the beta rays is measured with respect to the direction of nuclear polarization. If parity were conserved, equal numbers of electrons should be emitted parallel and antiparallel to the direction of nuclear polarization. The asymmetry found experimentally in the angular distribution proved that parity was not conserved in beta decay.

Let us now try to analyze this experiment using simple arguments and the diagrams of Fig. 1.1. For simplicity we assume that the nuclei are completely polarized. We choose the direction of nuclear polarization as our axis of quantization. The Co^{60} nuclei are therefore initially in a state whose angular momentum has a projection in the direction of our axis of $M_i = +5$, as is shown in Fig. 1.2. The final nuclear state has a total spin of 4. Its projection on the axis must be $M_f = +4$, as the electron and neutrino can each only carry an angular momentum of $\frac{1}{2}$, and angular momentum must be conserved. We see that the projections of the electron and neutrino angular momenta on our axis are determined by these considerations, and that the $m_e = m_\nu = +\frac{1}{2}$.

Knowing the directions of the lepton angular momenta selects four diagrams out of the sixteen of Fig. 1.1, corresponding to the four possible combinations of directions of polarization of the two particles. From these diagrams, for $m_e = m_\nu = +\frac{1}{2}$ in Fig. 1.1, we see that the electron goes 'up' if it is right-handed,

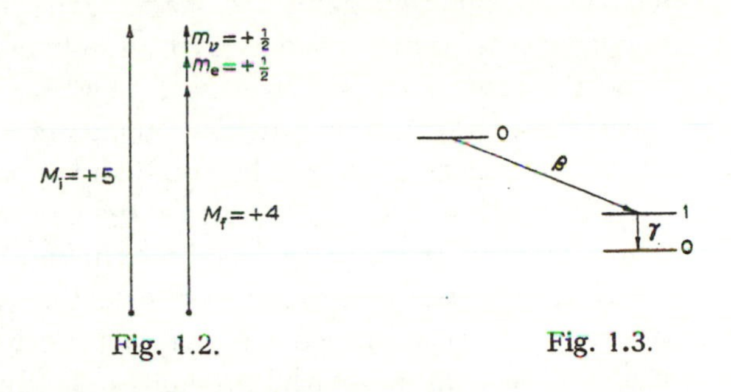

Fig. 1.2. Fig. 1.3.

or 'down' if it is left-handed, independent of the state of the neutrino. (If the electrons are unpolarized, as would be the case if parity *were* conserved, then half of them go up and half down, and there is no asymmetry.)

This experiment is thus completely equivalent to a measurement of the *degree of longitudinal polarization* of the beta particles emitted from Co^{60}. The relation between the results of the angular distribution experiment and the polarization measurement is thus *independent of beta decay theory*. The results of one of these experiments can only confirm or contradict the information obtained from the results of the other. They *cannot give any new additional information* about beta decay.

2. *Beta-gamma circularly polarized correlation*. Let us consider a decay scheme such as is shown in Fig. 1.3 in which a nucleus of spin zero decays by beta emission to a nucleus having spin one, which then decays by gamma emission to a nucleus having spin zero. The gamma ray is detected through a polarization analyzer, such as scattering in magnetized iron, in coincidence with the beta rays. The angular distribution of the beta rays is measured with respect to the direction of emission of the polarized gamma ray.

Let us take as our axis of quantization the direction of the *polarization* of the gamma ray (in the direction of emission if the gamma ray is *right-handed*, opposite if it is *left-handed*). The projection of the gamma ray angular momentum on our axis is thus $M_\gamma = +1$ as shown in Fig. 1.4. Since both the initial and final states of the nuclei in this beta-gamma cascade have spin zero, the total angular momentum carried away by the two leptons and the gamma ray must be zero as well. The projections of the lepton angular momenta on our fixed axis must therefore be equal and opposite to the projection of the gamma ray angular momentum. Thus $m_e = m_\nu = -\frac{1}{2}$, as shown in Fig. 1.4.

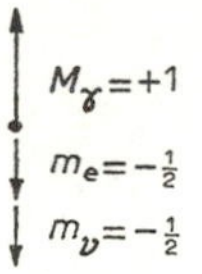

$M_\gamma = +1$
$m_e = -\frac{1}{2}$
$m_\nu = -\frac{1}{2}$

Fig. 1.4.

Now that the directions of the lepton angular momenta are known, we can immediately predict the direction of emission of the electron, as in the previous case. Since we have chosen our axis here so that $m = -\frac{1}{2}$, instead of $+\frac{1}{2}$, the predictions are opposite to those of the previous case. The electron goes *up* if it is *left*-handed, *down* if it is *right*-handed.

We see that this experiment is equivalent to the previous experiment and also measures the degree of longitudinal polarization of the electron.

This result can be expressed more conveniently if we note that the direction of emission of the gamma ray is *up*, if the gamma ray is right-handed, and *down* if it is left-handed. This is simply because we have chosen an axis in the direction of *polarization* of the gamma ray. The relation between the relative directions of emission and polarization of the electron and gamma ray can be summarized as follows: the electron and gamma ray are emitted in *opposite* directions if the senses of polarization of the two are the same (i.e. both left-handed or both right-handed); they are emitted in the *same* direction if the senses of polarization of the two are opposite.

The experimental result can be obtained by measuring the difference in the coincidence counting rate for two directions of beta ray emission, with a given direction of gamma ray polarization selected. It can also be obtained by looking at the change in counting rate with reversal of the direction of polarization of the gamma ray selected for a given direction of beta and gamma ray emission. The latter is more practical, since it involves only the reversal of a magnetic field and no changes in geometry.

3. *Electron-neutrino angular correlation*. Let us consider a case in which the initial and final nuclear states in a beta decay both have spin zero, and the angular correlation between the directions of emission of the electron and neutrino is measured. Since the total angular momentum carried by the leptons must be zero to conserve angular momentum, the components of the two angular momenta on any axis must be equal and opposite. One is $+\frac{1}{2}$, the other $-\frac{1}{2}$. The relevant diagrams in Fig. 1.1 are thus eight on the right-hand side of the figure. From the diagrams we see that if both leptons are right-handed, one goes up and the other down, and the same if both are left-handed. On the other hand, if one is left-handed and the other right-handed, they both go in the same direction.

The electron-neutrino angular correlation experiment therefore measures whether the longitudinal polarization or helicity of the electron has the *same* sense as that of the neutrino, or whether it has the *opposite* sense.

1.4. THE PHYSICAL BASIS OF THE SIMPLE PICTURE

In the three simple cases discussed above, we have been able to relate the results of angular distribution measurements to the angular momenta and polarizations of the particles in a manner independent of beta decay theory. Although the treatment is oversimplified, the underlying physical basis is valid. The existence of the diagrams of Fig. 1.1 and the selection and interpretation of those diagrams relevant to a particular decay follow directly from simple principles. A rigorous treatment can be based on these same principles to give exact relations between experimentally observed quantities which are independent of beta decay theory.

The essential feature underlying the interpretation of the diagrams of Fig. 1.1 is the simple, almost trivial, relation between momentum, angular momentum and sense of polarization. This interpretation is oversimplified, as is indicated by the representation of leptons as always going either 'up' or 'down'.

We know that there is no limitation in practice to these two directions, and that there should really be a continuous distribution. This inadequacy has nothing to do with beta decay theory, it results merely from our improper treatment of angular momenta in a quantum-mechanical problem. This can be rectified by a proper treatment of the angular momenta which gives the exact relations for all angular distributions. Because of the simplicity of the cases relevant in allowed beta decay, Racah algebra is not essential and we shall obtain all angular distributions without its use.

The law of conservation of angular momentum allows us to select the particular diagrams relevant to a particular decay. We have chosen particularly simple decay schemes for all our examples in the simple treatment. The angular momenta of the nuclear states were always chosen in such a way as to make possible a unique prediction of the projection on the axis of quantization of the total angular momentum carried by the leptons. In general this is not possible and the leptons can carry different angular momenta with different probabilities. This additional complication can also be handled by a proper treatment of angular momenta.

The essential physical feature underlying the existence of the diagrams of Fig. 1.1 is that *these sixteen diagrams represent all possible states for the electron and neutrino in allowed beta decay*. That there are only sixteen states results from 1) the restriction in allowed decay to lepton states of total angular momentum one-half, 2) the spin one-half of the electron and neutrino, and 3) the quantization of angular momentum in quantum mechanics. Except for the definition of what is meant by 'allowed decay', this is completely independent of beta decay theory.

Since the sixteen diagrams of Fig. 1.1 represent all possible lepton states, all meaningful information about allowed beta decay can be expressed by the answer to the following key question: *which of the sixteen states represented in Fig. 1.1 actually occur in allowed beta decay, and to what extent is each present?* If we know the answer to this question, we know everything about allowed beta decay required for the predictions of all experimental results. Conversely, the goal of beta decay experiments must be to accumulate sufficient experimental data to give a unique answer to the key question. Once such data has been acquired, no new information can be obtained from further experiments. They only check the previous results. The role of beta decay theory is then to provide a theoretical answer to the key question which can be compared with the answer given by experiment. Once we have obtained the answer to the key question from any theory, all experimental predictions follow automatically and there is nothing new which theory can tell us about allowed beta decay.

It is convenient to express this key question in the language commonly used in discussing nuclear reactions. We can say that each of the diagrams in Fig. 1.1 represents a *channel* for the electron and neutrino emitted in beta decay. The final state of the leptons after the decay can be expanded in a set of 'partial waves', each corresponding to a given channel. The amplitude of each partial wave represents the probability amplitude for the decay to occur in that particular channel. If the amplitudes for all sixteen channels are known, this specifies the state of the electron and neutrino completely, and the results of all measurements made on this state can be predicted without further information.

Since we are treating a quantum-mechanical system, the probability amplitudes are complex. The relative magnitudes indicate the relative probabilities for decay into the various channels. The relative phases are significant in predicting the results of experiments in which different channels can contribute coherently and interfere.

1.5. THE GENERAL PROGRAM FOR A RIGOROUS TREATMENT

A rigorous formulation of the simple arguments presented in this chapter should be based upon the principles discussed in the preceding section:

1. Description of the final state in terms of partial waves.
2. Conservation of angular momentum.
3. Restriction of the possible lepton states to a small number as a result of a) the quantization of angular momentum in quantum mechanics, b) spin one-half of electron and neutrino, c) restrictions of the possible values of lepton angular momentum (e.g. to $j = \frac{1}{2}$ in allowed decay).
4. The relation between angular momentum, linear momentum and helicity.

This is presented in the remaining chapters of this book in the following stages:

1. A rigorous specification is given for the electron-neutrino state. The experimentally measurable quantities are examined. The relation between the partial wave amplitudes, experimental measurements and fundamental theory is discussed.
2. The rigorous formulation is applied to allowed beta decay. A complete description of all experimental results is given in terms of the partial wave amplitudes. This description is oversimplified, like the diagrams of Fig. 1.1, describing angular distributions only qualitatively. Except for this inaccuracy all physical aspects of the experiments are treated properly.
3. Development of the properties of angular momentum necessary for the proper exact treatment of the angular distributions. This is mainly the quantum-mechanical version of the simple argument that a 'right-handed particle whose angular momentum is directed upwards must be going up'.
4. Application of the results on angular momentum to give the correct angular distributions for all experiments.

CHAPTER 2

SPECIFICATION OF THE INITIAL AND FINAL STATES

2.1. GENERAL

We now begin to develop a rigorous and precise formulation of the simple arguments of the preceding chapter based on the principles enumerated in section 5. We begin with the description of the final state in terms of partial waves. Let us assume that the system is prepared in a certain way; i.e. that it is in a certain initial state, before the beta decay takes place. We wish to examine all the possible final states (channels) and to find the relation between the amplitudes of the different partial waves and the results of experimental measurements performed on the system. We *do not consider* the relation between the final state partial wave amplitudes *and the initial state*: this is the task of beta decay theory.

The first step is to specify the states; i.e. to label the channels. This requires a set of quantum numbers sufficiently complete to describe those features of the initial and final states which are simply related to beta decay. We shall choose these quantum numbers to be the eigenvalues of a particularly simple set of commuting dynamical variables. The results of the usual beta decay experiments can then be shown to be given by expectation values of simple functions of these variables. These relations expressing the experimental results as expectation values of certain variables will then be independent of beta decay theory. They will depend only upon the validity of non-relativistic quantum mechanics and translational and rotational invariance (conservation of linear and angular momentum).

2.2. SPECIFICATION OF LEPTON STATES

The electron and neutrino are elementary particles of spin one-half. An electron or neutrino state is specified completely by four quantum numbers, three describing the three spatial degrees of freedom and one describing the orientation of the spin. A total of eight quantum numbers is thus necessary for a complete description of the combined electron-neutrino state. The three spatial quantum numbers most suitable for our purposes are the magnitude of the linear momentum, p; the magnitude of the total angular momentum, j, and the projection of the total angular momentum on some fixed axis, m.

The orientation of the spin $\boldsymbol{\sigma}$ may be specified in several ways, as has been shown in Chapter 1, § 2. The most convenient for our purposes is the projection $(\boldsymbol{\sigma} \cdot \boldsymbol{p})$ of the spin on the *linear momentum*, $\boldsymbol{p}$, of the particle. This specification is usually called 'helicity', h. A particle is said to be in a state of *positive* helicity, $h = +1$, when the projection of its spin in the direction of the momentum $\boldsymbol{p}$ is $+\frac{1}{2}$. A particle in such a state is called *right-handed*. A particle is in a state of *negative* helicity, $h = -1$, and is *left-handed*, when the projection of its spin in the direction of its momentum is $-\frac{1}{2}$. The subscripts R and L will be used to denote *right* and *left*-handed respectively. Thus

$$h = \frac{\boldsymbol{\sigma} \cdot \boldsymbol{p}}{|p|}, \tag{2.1a}$$

$$h\psi_{\mathrm{R}} = +\psi_{\mathrm{R}}, \tag{2.1b}$$

$$h\psi_{\mathrm{L}} = -\psi_{\mathrm{L}}. \tag{2.1c}$$

The helicity can also be expressed in terms of the projection of the total angular momentum $\boldsymbol{j}$ on the linear momentum $\boldsymbol{p}$. Since the orbital angular momentum $\boldsymbol{l}$ is always normal to the linear momentum $\boldsymbol{p}$,

$$\boldsymbol{l} \cdot \boldsymbol{p} = 0$$

and

$$\boldsymbol{j} \cdot \boldsymbol{p} = (\boldsymbol{l} + \tfrac{1}{2}\boldsymbol{\sigma}) \cdot \boldsymbol{p} = \tfrac{1}{2}\boldsymbol{\sigma} \cdot \boldsymbol{p},$$

thus

$$h = \frac{2\boldsymbol{j} \cdot \boldsymbol{p}}{|p|}. \tag{2.2}$$

We specify the state of each lepton by the four quantum numbers (p, j, m, h). The eight quantum numbers describing the lepton states are thus

$$(p_e, j_e, m_e, h_e, p_\nu, j_\nu, m_\nu, h_\nu) , \tag{2.3a}$$

where the subscripts e and ν refer to the electron and neutrino respectively. The state of the leptons emitted in beta decay can then be specified completely in terms of a set of complex amplitudes

$$a(p_e, j_e, m_e, h_e, p_\nu, j_\nu, m_\nu, h_\nu) , \tag{2.3b}$$

defined for all possible combinations of eigenvalues of these eight quantum numbers.

One can ask whether these quantum numbers really exhaust the total number of degrees of freedom and give a complete description of the lepton state. There may be other internal degrees of freedom in addition to the spin. This is of particular interest since the Dirac equation describes leptons using four component spinors, having two spin states and two charge states. For the electron we need not worry about the charge degree of freedom. The law of conservation of charge prevents the emission of both positive and negative electrons in the same decay process. As long as we are considering only one decay at a time, only one charge state is possible for the electron in each decay. For the neutrino, however, the two charge states (the so-called neutrino and anti-neutrino) might well both be emitted in the same decay with different probabilities. We leave this question for later consideration, however, and assume for the moment that the eight quantum numbers (2.3a) are sufficient to specify the lepton state completely.

The specification of the complex amplitudes (2.3b) for all values of the eight quantum numbers thus gives all the information available for the prediction of the result of an arbitrary measurement on the electron and neutrino emitted in beta decay. Using the language of nuclear reactions, we can say that each set of possible eigenvalues for the eight quantum numbers defines a 'channel' for the leptons emitted in beta decay. Each channel has a probability amplitude (2.3b) for the partial wave describing decay into this particular channel.

2.3. SPECIFICATION OF NUCLEAR STATES

Except in the case of the neutron decay, the specification of initial or final states of nuclei undergoing beta decay requires the solution, either theoretically or experimentally, of a complicated many-body problem. As this is not feasible, nuclear states are not specified completely for beta decay. The spin and parity are the only nuclear properties directly relevant to beta decay whose measurement is feasible. It is customary to classify beta decay transitions according to the change in spin and parity between initial and final nuclear states. More detailed properties of complex nuclei are not measured independently; rather the experimental data from beta decay is interpreted using beta decay theory to obtain information about nuclear structure.

We therefore do not specify the nuclear states in terms of quantum numbers, as we do for the leptons. Instead, we consider the lepton amplitudes (2.3b) as defined independently for each nuclear transition. The relation between corresponding amplitudes for different cases should give information regarding the structure of the nuclei involved, but this is beyond the scope of our treatment.

The beta decay of the neutron is the only case giving unambiguous information about the beta decay process independent of assumptions regarding nuclear structure.

2.4. CONSERVATION LAWS AND GEOMETRICAL FACTORS

In the preceding sections we have considered the first of the four physical principles enumerated at the beginning of Chapter 1, § 5; namely the description of the final state in terms of partial waves. This principle makes possible the description of the lepton states using the diagrams of Fig. 1.1. We now consider the next principle, the use of conservation laws.

The complex amplitudes

$$a(p_e, j_e, m_e, h_e, p_\nu, j_\nu, m_\nu, h_\nu) \tag{2.3b}$$

depend upon eight independent quantum numbers. This formulation is completely general, valid in any co-ordinate system. It is of interest to separate the dependence of the amplitudes (2.3b) upon the eight quantum numbers into a 'physical' part, which depends on the physical nature of the beta decay process, and a 'geometrical part' which depends upon the position and orientation of the co-ordinate system. We shall then find that the dependence of the amplitudes (2.3b) upon the eight quantum numbers is severely restricted by the conservation laws.

We first note that without loss of generality we can choose a co-ordinate system in which the nucleus is at rest before the decay. This corresponds to the laboratory system in all practical cases. We shall then see that in this system the variables p_e and p_ν are not independent; they are related uniquely by the conservation laws of energy and momentum. The amplitudes (2.3b) must vanish for all values of p_e and p_ν which are not consistent with the conservation laws. Thus only one of these quantum numbers is independent and the number of independent quantum numbers is reduced to seven.

We next note that the conservation of angular momentum places similar restrictions on the values of the angular momentum quantum numbers. The treatment of angular momentum is more complicated than that of linear momentum because the different components of the angular momentum operator do not commute with one another. We can only specify the magnitude of a given angular momentum operator and its projection on *one* fixed axis. The form of the restrictions imposed upon the amplitudes (2.3b) by the angular momentum conservation law depends upon how this fixed axis is chosen; i.e. whether it is the direction of a nuclear polarization, the direction of emission of the neutrino, etc. Which direction is most convenient for the *orientation* of this axis depends upon the particular experiment under consideration. (This is in contrast with the choice of *position* and *velocity* of the co-ordinate system, where the laboratory system is clearly the best for all experiments). We therefore wish to retain a general formulation which can be valid for all possible orientations of the co-ordinate system. It is advantageous to choose angular momentum quantum numbers in such a way as to separate the *physical* and *geometrical* factors. The variation of the amplitudes (2.3b) with the quantum numbers m_e and m_ν depends both on the nature of the beta decay process and on the orientation of the co-ordinate system. Instead of m_e and m_ν we shall choose two different quantum numbers, a 'physical' quantum number which describes an intrinsic property of the system independent of the co-ordinate axes, and a 'geometrical' quantum number which merely specifies the orientation of the system with respect to the co-ordinate axes. We shall then find that for a given experiment and choice of co-ordinate axes, the dependence of the amplitudes upon the 'geometrical' quantum number is determined uniquely by angular momentum conservation, while only the 'physical' quantum number is relevant to the actual beta decay process. Thus the number of independent quantum numbers *of physical interest* is reduced to six.

If parity were conserved in beta decay, we could reduce the number of independent quantum numbers still further in the same way. The variation of the amplitudes (2.3b) with the two helicity quantum numbers, h_e and h_ν, depends both on the nature of the beta decay process and upon whether the co-ordinate system is left-handed or right-handed. We could again choose two different quantum numbers, such that one would describe an 'intrinsic' property of the system, independent of whether the co-ordinate system is left or right-handed, while the other could be the parity of the state and express the transformation property of the lepton state under space inversion. The dependence of the lepton amplitude upon the parity quantum number would then be determined completely by the parity conservation law and the total number of independent quantum numbers would be reduced to five.

Parity is not conserved in beta decay; however it is still of interest to replace the two helicity quantum numbers by an 'intrinsic' quantum number and one which expresses the transformation property of the state under space inversion. The latter quantum number is no longer trivial, and can not be disregarded as in the case of parity conservation. Rather it has important physical significance and the dependence of the lepton amplitudes upon it expresses the degree of parity non-conservation in beta decay.

2.5. A MORE CONVENIENT SPECIFICATION OF LEPTON STATES

We now consider the explicit application of the general arguments of the preceding section. Let us consider the three conservation laws individually in detail.

The law of conservation of momentum requires that the sum of the momenta of the electron, the neutrino and the recoiling nucleus be equal to zero in the system where the nucleus is initially at rest (the laboratory system). The law of conservation of energy requires that the sum of the kinetic energies of these three bodies be equal to a constant characteristic of the particular decay; namely the difference in internal energy of the initial and final nuclear states, with suitable corrections for the rest energy of the emitted electron†.

The restrictions on the values of the electron and neutrino

† If the energy difference between the nuclear states is calculated using masses of *bare nuclei* and the relation $E = Mc^2$, the rest energy of the electron must be subtracted in the cases of negaton and positon emission to give the total energy released as kinetic energy in the decay. For electron capture, no correction for rest energy is required, but the binding energy of the electron in its atomic orbit must be subtracted.

However, the masses of *bare* nuclei are generally not used in these calculations; it is conventional to use *atomic*, rather than nuclear masses, including the rest energy of the extranuclear electrons. In such a calculation, the rest mass of the electron is automatically taken into account in negaton decay, but twice the electron rest energy must be subtracted in positon decay. The calculation for electron capture is the same as with bare nuclear masses.

momenta imposed by these conservation laws are expressible in a simple approximate manner, because the kinetic energy of the recoil nucleus is small compared to all other energies. Since momentum conservation requires that the momenta carried by the leptons and by the recoil nucleus all be of the same order of magnitude, the kinetic energies carried by these particles (non-relativistically) is inversely proportional to their masses. The energy carried by a nucleus of mass number A is of the order of $(1/1840A)$ times the energy carried by a non-relativistic electron in the same decay. Relativistic corrections at energies commonly encountered in beta decay (about 1 MeV) do not change the order of magnitude. We can therefore neglect this energy in comparison with the lepton energies and require that the sum of the electron and neutrino energies be constant. There is no further restriction on the lepton momenta due to momentum conservation, as the recoil nucleus can take up the required momentum with negligible energy.

As an example let us consider the case where the kinetic energy of the leptons is twice the rest energy of the electron, $2mc^2$, or about one MeV. The momentum taken by the recoil nucleus varies from zero, for the case where the electron and neutrino have equal and opposite momenta, to a maximum value in the case where the neutrino is emitted with zero energy and the electron has the full kinetic energy, $T = 2mc^2$. The momentum of the electron is then

$$p_e = \frac{1}{c}\sqrt{(T + mc^2)^2 - (mc^2)^2} = \sqrt{8}mc\,.$$

The momentum of the recoil nucleus is equal and opposite to that of the electron. The kinetic energy of the recoil nucleus is

$$T_r = \frac{p_e^2}{2M} = 2\frac{m}{M}(2mc^2)\,,$$

where M is the mass of the nucleus. For the neutron decay, (m/M) is 1/1840, and the maximum kinetic energy carried by the recoil proton is 1/920 times the lepton energy. In decays of complex nuclei the maximum kinetic energy of the recoil nucleus is even smaller by one or two orders of magnitude and thus ranges from several hundred to only a few electronvolts.

Since the electron and neutrino energies are related, the two quantum numbers p_e and p_ν are not independent. The lepton amplitudes (2.3b) must vanish for all values of p_e and p_ν which do not satisfy the conservation laws. It is thus sufficient to specify only one of these quantum numbers. The customary choice is the electron momentum. This is usually expressed as the electron velocity, v_e, because many effects in beta decay are proportional to v_e/c.

The law of conservation of angular momentum requires that the total angular momentum carried by the leptons be equal to the difference between the angular momenta of the initial and final nuclear states. It is the total lepton angular momentum which is restricted by the conservation law, rather than the orientation m_e and m_ν of the individual lepton angular momenta. The consequences of the conservation law are thus more easily displayed if the total angular momentum of the two leptons $\boldsymbol{j}_t$, and its projection m_t on the fixed axis, are used in our basic set of quantum numbers:

$$\boldsymbol{j}_t = \boldsymbol{j}_e + \boldsymbol{j}_\nu \tag{2.4a}$$

$$m_t = m_e + m_\nu . \tag{2.4b}$$

The law of conservation of angular momentum restricts the possible values of j_t, the magnitude of the *total* lepton angular momentum, to those between the sum and the difference of the spins J_i and J_f of the initial and final nuclear states,

$$|J_i - J_f| \leqslant j_t \leqslant J_i + J_f . \tag{2.5}$$

An important example of this restriction is the case $J_i = J_f = 0$, for which j_t can only be zero.

Another restriction is that imposed on the quantum number m_t, the projection of $\boldsymbol{j}_t$ on the fixed axis. If the projections, M_i and M_f of the angular momenta of the initial and final nuclear states are known, then m_t is determined uniquely,

$$m_t = M_i - M_f . \tag{2.6}$$

Since $m_t = m_e + m_\nu$, we see that the quantum numbers m_e and m_ν are not independent, and the lepton amplitudes (2.3b) must vanish for all values of m_e and m_ν which do not satisfy the conservation law (2.6). The most convenient way to express this restriction is to choose j_t and m_t to replace the quantum numbers m_e and m_ν in our basic set specifying the amplitudes (2.3b). Then the dependence of the amplitudes upon the quantum number m_t is determined completely by the conservation law (2.6); they vanish for all values of m_t except that given by (2.6).

In all beta decay experiments in which measurements are made on the leptons, it is always possible in principle to measure M_i and M_f as well. A quantity which does not commute with M_i and M_f is never measured. Thus, in any experiment in which M_i or M_f is not measured, one need know only the probability distribution for M_i and M_f and consider each set of possible values separately. The contributions from different values are not coherent and there are no interference effects. If the probability distribution for M_i and M_f is known, the distribution for m_t is given by (2.6).

The probability distributions for M_i, M_f and m_t can be determined for any experiment with the aid of the principle of rotational invariance. There is no preferred direction in space, other than those defined specifically by the geometry of the experiment. All results must be independent of the orientation of the system with respect to the co-ordinate axes. Thus if measurements are made in decays from unpolarized nuclei, and no measurements are made to determine the polarization of the final nuclear state, then all possible values of M_i are equally probable, and the same is true for M_f and for m_t. If the nuclei are initially polarized, but no measurement is made on the final nuclear state, then M_i is determined uniquely, but M_f and m_t are not. The probability distributions for M_f and m_t are then determined by the principle that all values of M_f and m_t are possible, subject to the constraint that the vectors $\boldsymbol{J}_i$, $\boldsymbol{J}_f$ and $\boldsymbol{j}_t$ must form a triangle, and that the projection of $\boldsymbol{J}_i$ on the axis of quantization must be M_i. This principle is applied quantitatively in Chapter 5. All other cases of partial polarization are treated in a similar way.

The choice of j_t and m_t as basic quantum numbers separates the 'physical' and 'geometrical' aspects of the angular momentum quantum numbers, as discussed in the previous section. The quantum number j_t is independent of the orientation of the co-ordinate system. Instead of specifying the orientation of the lepton angular momentum with respect to the fixed axis, it specifies the *relative* orientation of the two lepton angular momenta; i.e. how they add together to form a total angular momentum j_t.

The dependence of the lepton amplitudes upon j_t is a physical property of the beta decay process, independent of 'geometrical' factors like the orientations of the directions of nuclear polarization with respect to the co-ordinate axes. The dependence of the lepton amplitudes upon the quantum number m_t, on the other hand, is determined completely through (2.6) by the orientations of the nuclear states with respect to the co-ordinate axes. This is merely a property of the geometry of the particular experiment under consideration and is completely independent of the beta decay process.

Although the dependence of the lepton amplitudes upon the quantum number m_t may be important for the analysis of a particular experiment, this dependence is completely determined by geometrical considerations. The quantum number m_t can therefore be disregarded in discussing those properties of the lepton amplitudes which are directly relevant to the beta decay process: in discussing the relative amplitudes for the various decay channels, one simply notes that for each set of values of all the other quantum numbers, including j_t, there are $2j_t + 1$ channels, and that the relative amplitudes for these $2j_t + 1$ channels are determined by geometrical factors alone.

The law of conservation of parity is known to be violated in beta decay. The beta decay process is not invariant under space inversion. It is nevertheless convenient to choose a set of basic quantum numbers which displays parity conservation or non-conservation in a simple way. The two helicity quantum numbers, h_e and h_ν are not invariant under space inversion. The helicity of a state changes sign under a space inversion which interchanges left and right. If parity were conserved, the amplitudes (2.3b) would not change in magnitude under space inversion; i.e. if all helicities were reversed.

The consequences of parity conservation are not conveniently displayed in terms of the lepton amplitudes (2.3b) because *two* quantum numbers change under space inversion, h_e and h_ν. This is similar to the case of the quantum numbers m_e and m_ν which were not convenient for displaying angular momentum conservation because they both changed under rotations of the co-ordinate system. By analogy with the case of angular momentum, let us replace the two individual helicity quantum numbers by two new quantum numbers such that only *one* of them changes under space inversion while the other expresses an internal property of the system which is invariant under space inversion; i.e. it is independent of whether our co-ordinate system is left or right-handed.

By analogy with the case of angular momentum, let us look for an operator which describes the helicities of the electron and neutrino *relative to one another*, rather than relative to a particular kind of co-ordinate system. Such an operator is the *relative helicity*, h_{rel}, defined by

$$h_{rel} = h_e h_\nu \,. \tag{2.7}$$

$h_{rel} = +1$ when the two leptons have the *same helicity*; $h_{rel} = -1$ when they have *opposite helicity*. h_{rel} is invariant under space inversion, since both h_e and h_ν change sign. The relative helicity is thus an internal property of the system which is independent of whether the co-ordinate system is left or right-handed.

Let us choose the relative helicity h_{rel} as one of our basic quantum numbers. As the second helicity quantum number we choose the neutrino helicity, h_ν. This choice appears rather arbitrary at this point; one might have as well chosen the electron helicity or some more symmetric function of the two lepton variables, such as the parity of the state. We shall see later on that the choice of the neutrino helicity is the most appropriate for the description of beta decay.

Of our new quantum numbers, only the neutrino helicity h_ν is not invariant under space inversion. The dependence of the amplitudes (2.8) upon h_ν displays directly the conservation or non-conservation of parity in beta decay.

If parity were conserved, the magnitudes of the lepton amplitudes, expressed in terms of h_{rel} and h_ν should not depend upon the value of h_ν. The *parity* of the state would be given by the relative phase of the two corresponding amplitudes having opposite values for h_ν and the same values for all other quantum numbers. In this case, a more suitable quantum number than h_ν would be the *parity* π, of the state. Only one value of π would be allowed in any particular decay, determined by parity conservation and the parities of the initial and final nuclear states. The dependence of the lepton amplitudes upon π would be trivial, and its specification would not be relevant to beta decay. Thus if parity were conserved, we would require only five independent quantum numbers, v_e, j_e, j_ν, j_t, and h_{rel}. Note that the *relative* helicity remains a good choice for a basic quantum number even when parity *is* conserved.

Let us now write the lepton amplitudes in terms of our new set of six independent quantum numbers,

$$a(v_e, j_e, j_\nu, j_t, h_{\text{rel}}, h_\nu) \ . \tag{2.8}$$

The angular momentum and helicity quantum numbers have discrete eigenvalues whereas the electron velocity has a continuous spectrum. The two helicity quantum numbers each have only two possible values, and we shall see below that the angular momentum quantum numbers are limited to only a few possible values for any specific case. Thus the total number of possible combinations of angular momentum and helicity quantum numbers is relatively small, particularly for allowed transitions. It is therefore convenient to consider the lepton amplitudes (2.8) as a certain discrete number of independent complex amplitudes, one for each possible combination of eigenvalues for the angular momentum and helicity operators, with each amplitude being a function of the continuous variable v_e.

2.6. RESTRICTIONS ON THE VALUES OF THE ANGULAR MOMENTA

We now proceed to the third principle enumerated in Chapter 1, § 5. The values of the angular momentum quantum numbers are limited because the nuclear radius is small and the energies of leptons emitted in beta decay are relatively low. More precisely, the nuclear radius is small compared to the DeBroglie wave lengths of the leptons.

It is difficult for leptons to carry orbital angular momentum in beta decay. Beta decay energies are of the order of the rest energy of the electron, 0.5 MeV, and lepton momenta of at most several times mc occur in common transitions, where m is the rest mass of the electron and c is the velocity of light. A lepton of momentum mc can carry one unit $\hbar$ of orbital angular momentum if it is emitted classically at a distance

$$d = \frac{\hbar}{mc} \tag{2.9}$$

from the center of the nucleus. This is just the Compton wave length of the electron, 3.85×10^{-11} cm. Since nuclear radii are of the order of 10^{-12} cm, d is at least an order of magnitude larger than the nuclear radius. A lepton emitted classically from the nuclear surface would therefore carry only a tiny amount of orbital angular momentum, much less than $\hbar$. Quantum-mechanically this means that the *average* orbital angular momentum carried by the lepton is much less than $\hbar$; thus the probability that it carries orbital angular momentum is small. One can also describe this effect by saying that the 'centrifugal barrier' reduces the amplitude of a lepton wave function of high angular momentum at small radii. A wave function of orbital angular momentum l behaves as $(r/\lambda\!\!\!^{-})^l$ near the origin, where $\lambda\!\!\!^{-}$ is the DeBroglie wave length of the lepton. This is just equal to d in (2.9) and is therefore at least an order of magnitude larger than the nuclear radius. Thus $(r/\lambda\!\!\!^{-})$ is small and the probability of emission of leptons of orbital angular momentum l decreases very rapidly with l.

Beta decay transitions are commonly classified according to the angular momentum carried by the electron and neutrino. The allowed transitions are those in which the electron and neutrino each carry a minimum amount of angular momentum, $j_e = j_\nu = \frac{1}{2}$. Since the orbital angular momentum is not a constant of the motion for a relativistic particle, but only the total angular momentum j, we cannot require that the leptons be in an s-state with $l = 0$. The best we can do is $j = \frac{1}{2}$ which is a mixture of s- and p-states.

In the allowed transitions, there are therefore only eight independent amplitudes, (2.8), each depending upon the electron velocity, v_e. j_e and j_ν are each determined uniquely to be $\frac{1}{2}$. There are thus only two possible values for the total lepton angular momentum, j_t; namely 0 and 1. As there are two possibilities for each of the helicity quantum numbers, there are in all eight possible combinations of values of the angular momentum and helicity quantum numbers in allowed beta decay transitions.

The number of possible lepton states is reduced even further in the non-relativistic limit; i.e. when the velocity of the electron v_e approaches zero. (The neutrino is of course always relativistic.) The orbital angular momentum of the electron becomes a constant of the motion at $v_e = 0$. In this limit it is convenient to choose the *parity* of the electron state as a basic quantum number, rather than the electron helicity or relative helicity. At $v_e = 0$, the state $j = \frac{1}{2}$, even parity, becomes a pure s-state; the state of $j = \frac{1}{2}$, odd parity, becomes a pure p-state. Since the electron tends to be emitted with the minimum possible orbital angular momentum, only the s-state is allowed; the p-state is forbidden. There is thus only one allowed electron state in the non-relativistic limit, $v_e = 0$; namely the s-state which has a *definite parity*, and is a mixture of both helicities with equal probabilities. *The effects of parity non-conservation thus do not appear in measurements on electrons emitted at low velocities.* There are only four independent lepton states in allowed transitions at $v_e = 0$ instead of eight.

The reduction of the number of possible lepton states by a factor of two in the non-relativistic limit can also be expressed as a condition on the lepton amplitudes (2.8) expressed in terms of the helicity quantum numbers. At zero velocity, the electron is always emitted in a state which is an equal mixture of both helicities. The amplitudes for emission of electrons of zero velocity must therefore be of equal magnitude for both values of the electron helicity. Since the state has *positive* parity, the relative phase of the two equal amplitudes for emission of electrons with opposite helicity must be 0°.

Let us now state this condition precisely in terms of the amplitude (2.8) using the relative helicity quantum number. We note that for lepton states having the same neutrino helicity, the states of opposite *electron* helicity have opposite *relative* helicity. The condition then can be stated as follows:

Corresponding lepton channels, having opposite *relative helicity* and equal values for all other quantum numbers, have equal amplitudes in the limit where the electron velocity goes to zero.

$$\lim_{v_e \to 0} a(v_e, \tfrac{1}{2}, \tfrac{1}{2}, j_t, h_{rel} = +1, h_\nu) = \lim_{v_e \to 0} a(v_e, \tfrac{1}{2}, \tfrac{1}{2}, j_t, h_{rel} = -1, h_\nu) . \tag{2.10}$$

2.7. THE EXPERIMENTALLY MEASURABLE QUANTITIES

The complex amplitudes

$$a(v_e, j_e, j_\nu, j_t, h_{rel}, h_\nu) \tag{2.8}$$

give a complete specification of the leptons emitted in beta decay. The ideal experimental procedure for determining the nature of the beta decay process would be to measure these amplitudes directly. This is however not feasible because most of the dynamical variables involved cannot be measured directly.

The *neutrino* cannot be detected at all in a beta decay experiment. The detection of neutrinos in general is very difficult, requiring very large detectors and a high neutrino flux, such as that emitted from a nuclear reactor. The detection of neutrinos identified as coming from particular decays or from a particular small sample is out of the question at present. All measurements of properties of neutrinos emitted in beta decay must be indirect. Measurements of momenta and angular momenta of *both* the electron *and* the recoil nucleus must be used together with the conservation laws to obtain properties of the neutrino.

The *angular momentum of the electron* cannot be measured directly in any feasible way. It can be determined indirectly with the aid of the conservation laws in cases where knowledge of the angular momenta of the initial and final nuclear states determines uniquely angular momenta of both the electron and neutrino. The electron angular momentum can also be determined indirectly by angular distribution measurements.

The *helicity* of the electron is measurable directly, but only in a statistical way. The helicity of an *individual* beta ray cannot be measured in practice. All the methods for measuring beta ray polarization involve a scattering or radiation process for which only a statistical description is possible. The results are expressed as a probability depending upon the direction of electron polarization. Thus only the *mean* helicity of a *beam of electrons* can be measured.

The *velocity of the electron* is the only lepton variable which can be measured directly for an *individual* decay. It is possible to select beta rays within a definite energy region and to measure other properties of those decays in which electrons are emitted in this energy region.

Because all measurements except that of the electron velocity are statistical in nature, we cannot measure the lepton amplitudes (2.8) directly by holding all the quantum numbers fixed except one and seeing how the amplitudes vary as a function of this quantum number. We cannot select, for example, those decays in which right-handed electrons are emitted. We cannot measure the helicity of those neutrinos which are emitted in coincidence with right-handed electrons. We can only measure mean or expectation values of functions of the electron and neutrino helicities and other variables. A typical measurement gives the mean value of some function of the lepton helicity and angular momentum variables for a particular electron energy.

All measurements except those of the electron velocity and helicity are indirect, requiring either the use of conservation laws or the interpretation of an angular distribution measurement or both. Let us now examine in detail how each quantity can be measured indirectly.

The momentum of the neutrino is not one of our basic set of independent quantum numbers, (2.8). However, it is related to the neutrino angular momentum and helicity and is therefore an important quantity to measure. We have seen in the preceding section that the energy of the neutrino, and therefore the *magnitude* of its momentum, are determined from the electron energy to a very good approximation, using the energy conservation law and neglecting the kinetic energy of the recoil nucleus. The *direction* of the neutrino momentum, relative to that of the electron momentum, can be determined by measurements on the electron and recoil nucleus and using the conservation of momentum.

In practically all beta decay experiments, the nucleus is initially at rest in the laboratory system, which is therefore also the center-of-mass system. After the decay, the vector sum of the momenta of the electron, neutrino and recoil nucleus must add up to zero. If the electron energy is measured, then the magnitude of the electron momentum is known, and the magnitude of the neutrino momentum is known from energy conservation. The full momentum triangle can be determined by a measurement of either the *magnitude* of the recoil momentum or its direction. (In the latter case there is sometimes an ambiguity allowing for two possible solutions.) The angle between the electron and neutrino momenta is then determined, as shown in Fig. 2.1.

A simultaneous measurement of the electron energy and the recoil energy or direction is very difficult. Measurements on the recoil alone are usually made and then interpreted. The energy spectrum of the electrons is usually known, either theoretically or experimentally. Using this energy spectrum, one can calculate, for example, the energy spectrum of the recoil nuclei assum-

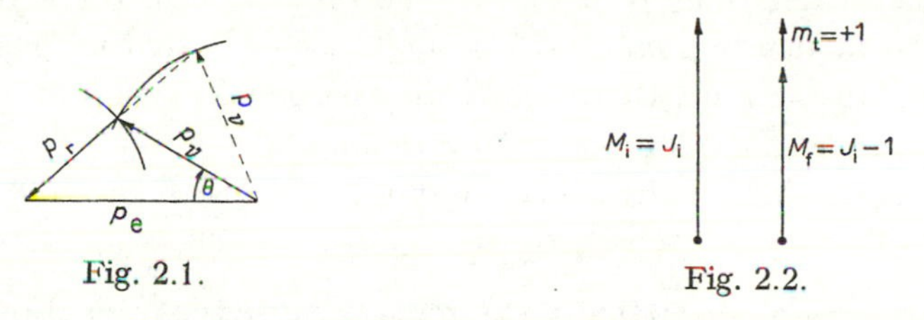

Fig. 2.1. Fig. 2.2.

ing some distribution for the electron-neutrino angular correlation. A comparison of the theoretical curve with experiment then tests the validity of the assumed electron-neutrino angular correlation. In particular we note that if the electron and neutrino tend to be emitted *parallel*, there is a large momentum transfer to the recoil nucleus and its energy spectrum is peaked at the high energy end. If the leptons tend to be emitted *anti-parallel* there is a low momentum transfer to the recoil and its spectrum is peaked at low energies. In section 5 we saw that recoil energies were very small: of the order of tens of electronvolts. Measurements of the recoil energy are therefore very difficult; they are easily perturbed by chemical, surface and molecular binding effects. The only successful direct recoil measurements have been made using noble gases.

The *lepton angular momenta* can sometimes be obtained from measurements of the angular momenta of the nuclear states and conservation of angular momentum, making use of the restriction of the lepton angular momenta to the smallest possible values consistent with the conservation law. Consider for example an allowed transition from a polarized nucleus where the spin of the initial nuclear state is one unit greater than the spin of the final state, J_f,

$$J_f = J_i - 1 \,.$$

The decay of Co^{60}, discussed in Chapter 1, § 3 is an example of such a transition.

Let the direction of polarization be taken as the axis of quantization, as shown in Fig. 2.2. Then the projection of the angular momentum of the initial state on the axis, M_i, is equal to J_i. The projection of the angular momentum of the final state, M_f, cannot be greater than $J_f = J_i - 1$. The projection of the lepton angular momentum m_t must be equal to $J_i - J_f$, by the conservation law. m_t must therefore be at least one. Since the transition is allowed the lepton angular momenta, j_e and j_ν are both equal to one-half. Since m_t is at least one, we have

$$m_e = m_\nu = +\tfrac{1}{2}\,.$$

This example is particularly simple because the projection of the final state angular momentum, M_f is determined uniquely by the polarization of the initial state and the requirement that the beta transition is allowed. The case $J_f = J_i + 1$ is more complicated, because M_f is not determined uniquely but can be $J_i + 1$, J_i, or $J_i - 1$. However, as discussed in section 5, the requirements of rotational invariance determine the relative probabilities of the three possibilities for M_f in a unique manner. This is treated in detail in Chapter 5. Thus the case used for our example represents in principle how the lepton angular momentum can be determined in experiments where the initial nuclear state is polarized.

A similar situation obtains in the case where the beta decay is followed by the emission of a gamma ray whose directions of emission and circular polarization are measured. Let us assume for simplicity the decay scheme as shown in Fig. 2.3. The initial state has spin zero, the intermediate state, after the beta decay has spin J, and the final state, after the emission of the gamma

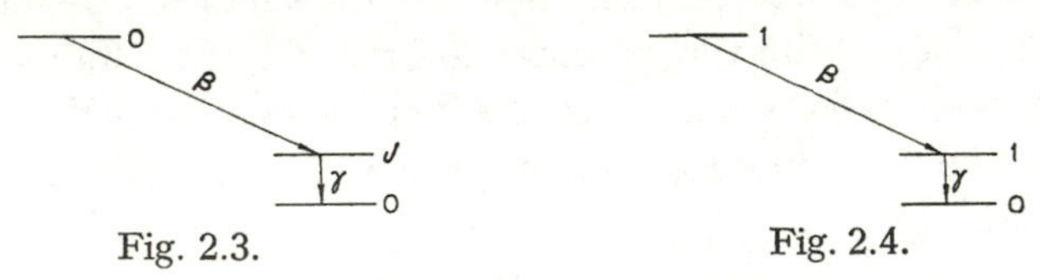

Fig. 2.3. Fig. 2.4.

ray, has spin zero again. Since both initial and final states have angular momentum zero, the sum of the angular momenta of all the emitted particles must also be zero. Thus the angular momentum of the leptons is equal and opposite to that of the gamma ray. If we take the direction of *polarization* of the gamma ray as our axis of quantization (either in the direction of emission or opposite, according to whether the gamma ray is right or left-handed) the projection of the angular momentum of the gamma ray on the axis is $+1$. The projection of the lepton angular momentum on this axis, m_t, is just equal and opposite. Thus $m_t = -1$ and is determined uniquely by the polarization measurement of the gamma ray.

For the case where $J = 1$ and the beta transition is allowed $j_e = j_\nu = \frac{1}{2}$ and $m_e = m_\nu = -\frac{1}{2}$.

Here again a particularly simple decay scheme has been chosen, in which m_t is uniquely determined. In the general case where several values of m_t are possible, the relative probabilities of the different values of m_t are also determined by the requirements of rotational invariance. This is treated in detail in Chapter 5.

Since both the neutrino momentum and the neutrino angular momentum can be determined by use of the conservation laws, the neutrino helicity, which is the projection of the angular momentum on the direction of the momentum, can in principle also be determined. In practice this is very difficult, since both the recoil and the polarization experiments are difficult individually. The combined polarization-recoil experiment was first performed, GOLDHABER [1958], using a beta-gamma cascade having a decay scheme like that of Fig. 2.3 with $J = 1$. The beta transition in this case is the capture of an orbital electron. The direction of the recoil was determined by measuring the Doppler shift of the gamma ray due to recoil motion, using resonance scattering to detect the shift. This case was particularly favorable, having a simple decay scheme with appropriate energies and a high resonance scattering. In general this method has not proved to be feasible (no other case as suitable as this one has been found), and more complicated indirect methods have been used to determine the helicity of the neutrino.

Measurements of angular correlations and angular distributions give indirect information regarding the angular momenta and helicities of the leptons. The relations between experiments measuring lepton angular distributions and the lepton channel amplitudes is discussed in detail in the following chapters.

The lepton angular momenta can sometimes be determined in a beta-gamma cascade by measuring the angular distribution of the *gamma ray* emitted after beta decay from polarized nuclei and using the angular momentum conservation law. Let us consider the beta-gamma cascade shown in Fig. 2.4, in which the initial and intermediate states both have spin one and the final state after gamma emission has spin zero. If the nuclei are polarized in the initial state and the direction of polarization is taken as axis of quantization, the projection of the angular momentum of the intermediate state M_{int}, can be either $+1$ or 0. (The beta decay is assumed to be allowed, so that the possibility of $M_{\text{int}} = -1$ is excluded.) The projection of the lepton angular momentum m_{t} is thus not uniquely determined, as in the case of Fig. 2.3, but can be either -1 or 0. These two cases can be distinguished by a measurement of the angular distribution of the *gamma ray*. If $m_{\text{t}} = 0$, then $M_{\text{int}} = +1$, and the projection of the gamma ray angular momentum $m_\gamma = +1$. If $m_{\text{t}} = -1$, $M_{\text{int}} = 0$, and $m_\gamma = 0$. The angular distributions for gamma rays having $m_\gamma = +1$ and $m_\gamma = 0$ are different. (Gamma rays having $m_\gamma = 0$, for example, cannot be emitted in the direction of the axis and their distribution favors the equatorial plane.) Thus the relative amounts of the two lepton angular momenta present can be determined from a measurement of the angular distribution of the gamma rays with respect to the direction of nuclear polarization.

2.8. THE RELATION BETWEEN LEPTON AMPLITUDES, EXPERIMENTAL MEASUREMENTS AND THEORY

A direct measurement of the lepton amplitudes (2.8) in a particular case of beta decay would yield all the information obtainable by experiment from that particular decay. In many cases the quantum numbers j_e and j_ν are restricted to their lowest possible values consistent with angular momentum conservation and the quantum number j_t is determined uniquely by angular momentum conservation. In such simple cases the only experimental information required is the dependence of the amplitudes (2.8) upon the two helicity quantum numbers as a function of the electron velocity. The mean electron helicity can be measured directly as a function of the electron velocity. However, the direct measurement of the neutrino helicity is not feasible. Indirect measurements, such as angular correlation experiments, must be used to determine the neutrino helicity. Such experiments measure the mean value of some complicated function of the neutrino helicity and other lepton variables.

In the general case, where the lepton angular momenta are not uniquely determined, the relative amplitudes of the different components cannot be measured directly. Complicated angular correlation and polarization experiments are required in this case to determine the dependence of the lepton amplitudes (2.8) upon the angular momentum quantum numbers, as well as the neutrino helicity. The results of these complicated experiments have generally been compared with theoretical predictions calculated directly from a fundamental theory of beta decay. These long calculations are really unnecessary, as all information concerning the experiments is contained in the amplitudes (2.8), which are more easily calculated from the theory. The complicated experiments can easily be interpreted in terms of our complete set of quantum numbers (2.8) without the use of beta decay theory.

The results of most of these experiments can be expressed as an average or expectation value of some function of the lepton variables. Since it is rare that more than two eigenvalues for a particular variable are relevant in any particular case, the average value gives directly the relative *magnitudes* of the amplitudes corresponding to the different eigenvalues.

Measurement of the *phases* of the complex lepton amplitudes requires some kind of interference effect. Only relative phases are measurable. If each of the complex amplitudes is multiplied by the same overall phase factor, no measurable effect is produced. To measure the interference between two lepton channels, it is necessary to measure a quantity *which does not commute* with the operators whose eigenvalues characterize the channels. Only in the measurement of such a quantity can the amplitudes for different channels contribute coherently and produce interference effects dependent upon the relative phases. Any quantity which commutes with all the basic operators whose eigenvalues specify the amplitudes (2.8) is a function of these basic operators. Its expectation or mean value in any experiment is simply the statistical average over all channels, with each channel having a statistical weight equal to the square of the magnitude of the corresponding amplitude.

2.9. INTERFERENCE EFFECTS

To measure interference between states of opposite helicity (longitudinal polarization), it is necessary to measure the projection of the spin on some axis other than the momentum (transverse polarization). This is the only measurement in which states of opposite helicity are mixed. For the neutrino such an experiment would be exceedingly difficult in practice, as it would require a direct measurement on the neutrino. Transverse polarization cannot be measured indirectly by use of conservation laws, because only the total angular momentum of the neutrino is measured in this way. The decomposition of this total angular momentum into spin and orbital angular momentum is possible only in the case of longitudinal polarization, since the orbital angular momentum has no component in the direction of the linear momentum.

We seem, however, to be spared the dilemma of the existence of a phase factor which is detectable in principle but not in practice. Experimental results indicate that the neutrino is always emitted in a state of 100% longitudinal polarization; i.e. in an eigenstate of the helicity. The amplitude for the emission of the neutrino with opposite helicity is therefore zero, and there is no interference effect.

For the electron a measurement of transverse polarization is feasible and can be used to detect interference between the amplitudes for emission with opposite helicity. This is not as important as one would expect. In the cases of greatest interest, allowed transitions at relativistic energies ($v_e \approx c$), the electrons have also been found experimentally to be 100% longitudinally polarized and there is no interference effect.

The most interesting cases of interference arise between the channels of different angular momentum, rather than helicity. The magnitude of the total lepton angular momentum, j_t, does not commute with the angular momenta of the individual leptons, $\boldsymbol{j}_e$ and $\boldsymbol{j}_\nu$. A measurement of the projection of the angular momentum of either the electron or the neutrino, m_e or m_ν, can contain coherent distributions from different values of j_t and thus measure the relative phases of these amplitudes.

CHAPTER 3

ALLOWED BETA DECAY IN WORDS AND PICTURES

3.1. THE SIXTEEN CHANNELS AND THEIR INTERPRETATION

In allowed beta decay the electron and neutrino are both emitted in a state of angular momentum $j_e = j_\nu = \frac{1}{2}$. Each particle has two possible values for the projection of the angular momentum on a fixed axis, $m_e = m_\nu = \pm\frac{1}{2}$ and two possible values for the helicity $h_e = h_\nu = \pm 1$. There are thus four possible angular momentum and helicity states for each particle, and sixteen possible states for the combined electron and neutrino system. For a given electron velocity v_e, there are sixteen possible channels for the decay, and the process is specified completely by the sixteen complex amplitudes for the partial waves for decay into these channels. In this section we do not consider the restriction on the amplitudes due to angular momentum conservation discussed in Chapter 2, § 6. This is taken into account in the following section.

Two types of measurements are made on leptons emitted in allowed beta decay. The first type are polarization experiments in which the mean helicities of particles are measured directly; the remaining group can be called '$(1 + A\cos\theta)$' experiments, in which the angular distribution of the emission of some particle is measured with respect to some axis defined by the directions of polarization or emission of other particles. The resulting distribution always has the form '$1 + A\cos\theta$' and the asymmetry parameter A gives information about the beta decay process. The exact relation between the parameter A and our basic set of quantum numbers is not immediately evident. We shall see that the value of A in any experiment can be expressed as a function of the mean values of some of the four quantum numbers m_e, m_ν, h_e and h_ν.

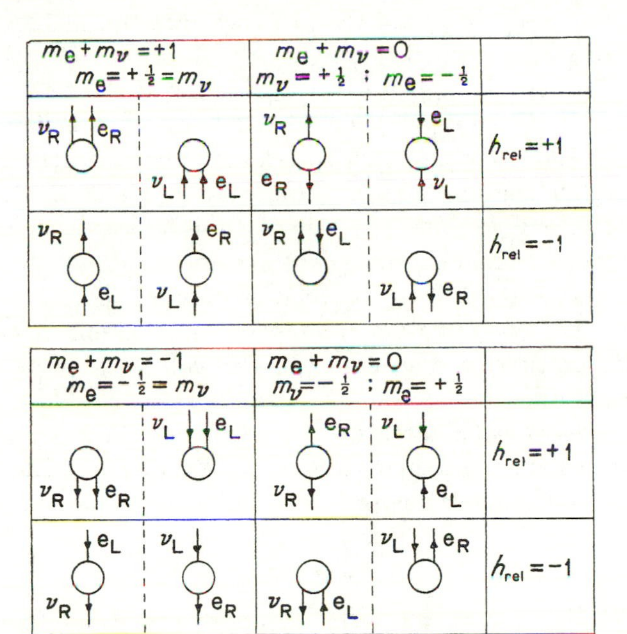

Fig. 3.1.

The various '$1 + A \cos\theta$' experiments can be interpreted with the aid of diagrams like those of Fig. 3.1, showing the sixteen possible channels for allowed decay. These are the same diagrams as in Fig. 1.1 with additional labels corresponding to the quantum numbers introduced in Chapter 2. In each diagram the lines indicate the paths of the electron and neutrino emitted from the nucleus, and the arrows indicate the direction of the spin of the particle. Each path is labeled by the nature of the particle and by its helicity, as e_R, e_L, ν_R or ν_L. The helicity is of course already indicated by the direction of the spin arrow with respect to the direction of emission. It is convenient to group the diagrams according to the projection of the total lepton angular momentum, $m_t = m_e + m_\nu$, upon the fixed axis, and according to the relative helicity, $h_{rel} = h_e h_\nu$. The sixteen diagrams are also divided into two groups of eight according to the sign of m_ν. The upper eight diagrams have $m_\nu = +\frac{1}{2}$, the lower eight have $m_\nu = -\frac{1}{2}$. We note that to each diagram in the upper eight, there corresponds a diagram in the lower eight which is obtained from the upper one by a rotation of 180°; i.e., by reversing the signs of m_e and m_ν without changing the helicities.

From the diagrams of Fig. 3.1, we see that the helicity and angular momentum quantum numbers also determine the direction of emission of the particles. This is to be expected, since these quantum numbers constitute a complete set which specify all properties of the state. The relation between helicity, angular momentum, and direction of emission (momentum) is just the fourth principle of Chapter 1, § 5. This can be expressed more precisely as follows:

1. A *right-handed* particle is emitted in the direction *parallel* to its angular momentum, a *left-handed* particle is emitted *antiparallel* to its angular momentum. This can be summarized by the relation

$$\mathbf{e}_p = h\mathbf{e}_m \,, \tag{3.1}$$

where $\mathbf{e}_p$ is a unit vector in the direction of the momentum and $\mathbf{e}_m$ is a unit vector in the direction of the angular momentum.

2. The direction of emission of the electron relative to that of the neutrino is expressed by the relation

$$\mathbf{e}_{p_e} \cdot \mathbf{e}_{p_\nu} = h_e h_\nu \mathbf{e}_{m_e} \cdot \mathbf{e}_{m_\nu} = h_{rel}\, \mathbf{e}_m \,. \tag{3.2}$$

The electron and neutrino are emitted in the *same* direction if $m_e m_\nu h_e h_\nu$ is positive; they are emitted in opposite directions if $m_e m_\nu h_e h_\nu$ is negative. In other words, they are emitted in the same direction if h_{rel} is positive and m_e and m_ν have the same sign, or if h_{rel} is negative and m_e and m_ν have opposite signs. They are emitted in opposite directions if h_{rel} is negative and m_e and m_ν have the same sign, or if h_{rel} is positive and m_e and m_ν have opposite signs.

The diagrams indicate that particles must be emitted either 'up' or 'down', either 'parallel' or 'antiparallel' to the directions of their angular momenta or to one another. This cannot be really true. Angular momenta are restricted to discrete eigenvalues, but linear momenta are not, and we should expect continuous angular distributions for the momenta of the particles. In fact we expect the '$1 + A \cos \theta$' distribution which characterizes these experiments. The diagrams of Fig. 3.1 therefore present an oversimplified picture. We shall consider them as only qualitatively correct, and assume that a diagram showing a particle emitted upward really means a continuous angular distribution with an asymmetry preferring the upward direction. A rigorous treatment justifying this interpretation of the diagrams is given in Chapter 4.

From our two observations above, we can draw the following conclusions:

1. The measurement of the angular distribution of the direction of emission of a lepton measures the mean value of e_p, which is equal to the mean value of the product he_m. If m is known, this gives a measurement of the *helicity* of the lepton. The same is true of measurements of the angular distribution with respect to a direction of nuclear polarization or the direction of emission of a gamma ray, since the directions of nuclear polarization and gamma ray emission are related to that of the angular momentum of the lepton by conservation of angular momentum.

2. The measurement of the angular distribution of the electron with respect to the neutrino is a measurement of $m_e m_\nu h_{rel}$. If the relative direction of the two angular momenta is known, i.e., whether $m_e = +m_\nu$ or $-m_\nu$, this is a measurement of the mean relative helicity.

3.2. THE FERMI AND GAMOW-TELLER LEPTON STATES

The two conclusions of the preceding section are precise statements of the relation between momentum, angular momentum and helicity discussed in Chapter 1, §§ 4 and 5. These conclusions show that the results of the angular distribution '$1 + A \cos \theta$' experiments are expressed very simply in terms of the helicity and angular momentum quantum numbers of the individual lepton, h_e, h_ν, m_e, and m_ν. However, it was shown in Chapter 2, § 5 that the consequences of angular momentum conservation and rotational invariance are described more conveniently in terms of the total lepton angular momentum quantum numbers, j_t and m_t. The operator j_t does not commute with the individual lepton angular momenta m_e and m_ν. Thus a simple description of the relation between momentum, angular momentum and helicity seems to be incompatible with a simple description of angular momentum conservation. The (m_e, m_ν) representation is preferable for the former, the (j_t, m_t) representation for the latter.

We shall deal with this dilemma as follows: we define the lepton amplitudes (2.8) in the (j_t, m_t) representation. The dependence of the amplitudes upon j_t is then a property of the beta decay process, independent of the orientation of the system. The dependence of the amplitudes upon m_t is completely independent of the beta decay process and is determined by the orientation or polarization of the system and by angular momentum conservation. The dependence upon m_t is not a property of the beta decay process and is omitted in the expression (2.8). However, this dependence upon m_t must be determined from the kinematics of any particular experiment in order to predict angular distributions. For a given dependence of the lepton amplitudes upon j_t and m_t, the mean values of m_e and m_ν can be calculated. The angular distributions are then given in terms of the mean values of m_e, m_ν and the helicities, as was shown in the preceding section. The dilemma is thus reduced to the necessity of calculating the mean values of m_e and m_ν in the (j_t, m_t) representation.

Let us now examine the possible lepton channels in allowed transitions in the (j_t, m_t) representation. There are only two possible values for j_t; namely 0 and 1, since $j_e = j_\nu = \frac{1}{2}$. The transitions for which $j_t = 0$ are called allowed *Fermi* transitions; those for which $j_t = 1$ are called allowed *Gamow-Teller* transitions.

As a result of angular momentum conservation the value of j_t determines selection rules for the change in the angular momentum of the nucleus, as given by (2.5)

$$|J_i - J_f| \leqslant j_t \leqslant J_i + J_f \, . \tag{2.5}$$

Setting $j_t = 0$ or $j_t = 1$ we obtain the selection rules for allowed transitions. In allowed Fermi transitions, there can be no change in angular momentum of the nucleus. In allowed Gamow-Teller transitions the change in nuclear angular momentum can be either one unit or zero, but zero to zero transitions are excluded.

$\Delta J = |J_i - J_f| = 0$ *in allowed Fermi transitions*, $(j_t = 0)$.

$\Delta J = |J_i - J_f| = 0$ or 1 *in allowed Gamow-Teller transitions*, $(j_t = 1)$.
but $J_i = J_f = 0$ *excluded*

Let us now express the diagrams of Fig. 3.1 in the Fermi-Gamow-Teller classification. There are eight independent channels, corresponding to the different values of h_e, h_ν, and j_t. The different values of m_t need not be considered as the dependence of the channel amplitudes upon this quantum number is determined completely by kinematics and angular momentum conservation. Since the quantum numbers h_e, h_ν and j_t do not refer to the direction of any fixed axis in space, the directions 'up' and

Gamow-Teller $J_t=1$; $\Delta J=0;1$ $(0 \not\to 0)$	Fermi $J=0$; $\Delta J=0$	
ν_R ν_L $\frac{2}{3}\,(p_e \parallel p_\nu)$ $\frac{1}{3}\,(p_e \nparallel p_\nu)$	ν_R ν_L $p_e \nparallel p_\nu$	$h_{rel}=+1$
ν_R ν_L $\frac{2}{3}\,(p_e \nparallel p_\nu)$ $\frac{1}{3}\,(p_e \parallel p_\nu)$	ν_R ν_L $p_e \parallel p_\nu$	$h_{rel}=-1$

Fig. 3.2.

'down' in Fig. 3.1 do not have any relevance for this classification. Only the relative directions of motion of the electron and neutrino are significant.

The eight independent channels for allowed beta decay in the Fermi-Gamow-Teller classification are illustrated schematically in Fig. 3.2.

Since absolute directions have no meaning, the notation $p_e \parallel p_\nu$ or $p_e \nparallel p_\nu$ is used rather than drawing pictures to indicate whether the electron and neutrino are emitted parallel or antiparallel to one another. The helicity of the neutrino is indicated by the letter L or R. The relation between Figs. 3.1 and 3.2 is very simple. For the Fermi transitions $m_e + m_\nu$ is always zero and the diagrams from Fig. 3.1 can be transferred directly to Fig. 3.2. For Gamow-Teller transitions, $m_e + m_\nu$ is either $+1$, -1, or 0. In Fig. 3.1 we see that the relative directions of emission of the electron and neutrino are always the same for $m_t = +1$ and for $m_t = -1$, but are the reverse for $m_t = 0$.

The different states of m_t all have the same probability if the initial nucleus is unpolarized and all states of polarization of the final nucleus are weighted equally. The resulting electron-neutrino correlation is like the $m_t = \pm 1$ diagram in two thirds of the cases and like $m_t = 0$ in one third. (In special experiments where there is a selection of certain states of nuclear polarization, the different m_t states are not weighted equally. This effect must be taken into account in determining the angular correlation.)

Let us now re-examine the two conclusions of the previous section in the Fermi-Gamow-Teller classification.

1. The lepton angular distributions depend upon the angular momenta m of the individual leptons. For pure Fermi transitions, $j_t = 0$, and the lepton state must be spherically symmetric. The mean angular momenta of the individual leptons must also be zero. There is therefore no asymmetry in the angular distributions in pure Fermi transitions. For pure Gamow-Teller transitions, $j_t = 1$, and the mean value of m_t can be determined from angular momentum conservation in the relevant experiments. Since j_e and j_ν are parallel in Gamow-Teller transitions the mean values of m_e and m_ν are both equal to $\frac{1}{2}\langle m_t \rangle$. The experiment again measures the mean lepton helicity.

2. The electron-neutrino angular correlation is described directly by Fig. 3.2. Again the experiment measures the mean relative helicity h_{rel}. The relative direction of the two angular momenta is determined by the classification of the decay as Fermi or Gamow-Teller. An additional factor of $\frac{1}{3}$ appears in the Gamow-Teller case in the mean value $\langle m_e m_\nu \rangle$.

3.3. MIXED TRANSITIONS AND INTERFERENCE

The selection rules for Fermi and Gamow-Teller transitions are not mutually exclusive. They overlap for the case $J_i = J_f \neq 0$, which is allowed for both types of transitions. Such cases are called '*mixed transitions*'. The leptons emitted in mixed transitions can be either in $j_t = 0$ state, a $j_t = 1$ state or a mixture of the two. The relative magnitudes and phases of the Fermi and Gamow-Teller components are not determined a priori by any conservation law and must be determined by experiment.

The conclusions of the preceding section regarding angular distributions are given from Fig. 3.2 only if the Fermi and Gamow-Teller contributions are *incoherent* and *do not interfere*. In such a case, the results of any experiment is simply the sum of the results for the Fermi and Gamow-Teller components with each contribution weighted with the square of the amplitude for the corresponding channel.

As has been discussed in Chapter 2, § 6, interference between the Fermi and Gamow-Teller amplitudes can be expected in the measurement of a quantity which does not commute with j_t the variable characterizing the Fermi-Gamow-Teller classification. Such a quantity is the angular momentum of an individual lepton, m_e or m_ν. It is just these quantities which are measured in the experiments determining the angular distribution of individual leptons relative to a fixed axis, and one should expect to find interference effects in these experiments. Note, however, that in the *electron-neutrino* angular correlation experiment, the quantity measured is $\boldsymbol{p}_e \cdot \boldsymbol{p}_\nu$ which commutes with j_t. There should therefore be no interference effect in this experiment and the result is simply the sum of the individual Fermi and Gamow-Teller contributions.

The physical nature of the interference effect can be understood by examining the upper and lower sets of diagrams corresponding to $m_t = 0$ in Fig. 3.1. Both of these sets of diagrams correspond to the same change in nuclear angular momentum; namely $\Delta M = 0$. It is therefore possible to distinguish between them by measurements on the nuclear states. One can distinguish between an upper diagram and the corresponding lower diagram having the same helicities by measuring the direction of emission of either the electron or the neutrino with respect to a fixed axis. (The fixed axis would be determined by the direction of polarization of one of the nuclear states or of a subsequently emitted gamma ray.)

The Fermi amplitude and the $m_t = 0$ Gamow-Teller amplitude are both linear combinations of the two amplitudes represented by corresponding upper and lower diagrams in Fig. 3.1 for $m_t = 0$. Both diagrams contribute equally in magnitude for the two cases, but the relative phases are opposite for the Fermi (singlet) and Gamow-Teller (triplet) states. Conversely, the states represented by the individual $m_t = 0$ diagrams in Fig. 3.1 are each coherent linear combinations of the Fermi and Gamow-Teller states with different phases. A measurement which distinguishes between the upper and lower $m_t = 0$ diagrams measures the relative magnitude and phase of the Fermi and Gamow-Teller amplitudes.

Note that the detection of interference effects (i.e. the distinction between upper and lower diagrams in Fig. 3.1) requires the definition of a fixed axis independent of the directions of lepton emission. Such an axis would be the direction of a nuclear polarization or of emission of a polarized gamma ray. Such an axis is not defined in an experiment measuring the angular correlation between the electron and neutrino. Fig. 3.1 shows that this type of experiment cannot distinguish between the corresponding upper and lower diagrams. Both diagrams give exactly the same results for such an experiment. Thus we see again that no interference effects between the Fermi and the Gamow-Teller components occur in e–ν angular correlation.

Note also that adjacent diagrams in Fig. 3.1 having the same values of m_t and h_{rel} differ from one another only by having all helicities reversed, and therefore both directions of emission reversed. The two diagrams in such an adjacent set transform into one another under space inversion (reversal of all helicities). If parity were conserved, the amplitudes for the two adjacent diagrams would be of equal magnitude and no asymmetry would be observed in the angular distribution of either the electron or the neutrino with respect to a fixed axis. The asymmetries of the contributions from each of the two adjacent diagrams would be equal and opposite and would cancel one another. Thus if parity were conserved, interference between the Fermi and the Gamow-Teller amplitudes could not be measured by a lepton angular distribution experiment; an additional helicity measurement would be necessary to distinguish between the two adjacent diagrams.

The two conclusions of the preceding sections can be restated for mixed transitions as follows:

1. The lepton angular distribution depends upon the mean helicity h and the angular momentum m of the emitted lepton. The latter depends upon the relative magnitudes of the contributions from the corresponding upper and lower diagrams in Fig. 3.1. For the $m_t = 0$ case, this depends upon the *relative magnitude* and *phase* of the Fermi and Gamow-Teller components, for corresponding channels having the same helicity quantum numbers. Only the $m_t = 0$ channels are common to Fermi and Gamow-Teller transitions; thus $\langle m_e \rangle = -\langle m_\nu \rangle$ and the interference effects on the electron and neutrino angular distributions must be equal and opposite.

2. The electron-neutrino angular correlation is described by combining the appropriate Fermi and Gamow-Teller diagrams in Fig. 3.2 weighted by the relative probability of the two types of transitions (the squares of the corresponding amplitudes). The experiment again measures the mean relative helicity, if

this is the same for both the Fermi and Gamow-Teller components. If the mean relative helicity is different for the two components, the experiment measures a linear combination of the two relative helicities, weighted by the corresponding relative probabilities, with an additional weighting factor $-\frac{1}{3}$ for the Gamow-Teller component.

3.4. THE RELATION BETWEEN EXPERIMENTAL RESULTS AND THE SPECIFICATION OF LEPTON STATES

We can now summarize the conclusions of the preceding section in the case of the Fermi-Gamow-Teller classification and obtain relations between experimental results and the specification of the lepton states. First we shall express the result of each type of experiment in terms of the quantities appearing in the lepton amplitudes (2.8). Then we shall see how a complete set of experiments can be found to determine these amplitudes.

1. *Experiments measuring the angular distribution of leptons emitted in allowed transitions with respect to a direction of nuclear polarization or polarized gamma ray emission.* We have seen that in such an experiment the mean value of the product $\langle he_m \rangle$ is measured.

a. Pure Gamow-Teller transitions. The angular momentum of the leptons m_t is known from angular momentum conservation; the experiment therefore measures the *helicity* of the lepton whose angular distribution is measured.

b. Pure Fermi transitions. The angular momentum of the leptons $j_t = m_t = 0$, and corresponding upper and lower diagrams in Fig. 2.1 contribute equally. There is therefore no effect.

c. Mixed transitions. If the angular momentum of the emitted lepton is known, the experiment measures the helicity. In general the angular momentum of the lepton is not known. It depends upon the relative magnitudes and phase of the Fermi and Gamow-Teller amplitudes and is not determined by angular momentum conservation as in the case of the pure Gamow-Teller transition. The experimental result therefore gives a function of the lepton helicity and the relative magnitudes and phases of the Fermi and Gamow-Teller amplitudes.

2. *Experiments measuring the angular correlation between the direction of emission of the electron and neutrino.* We have seen that in such an experiment the mean value of the product $\langle m_e m_\nu h_{rel} \rangle$ is measured.

a. Pure Fermi transitions. The total lepton angular momentum $j_t = m_t = 0$ and the relative directions of emission are given by Figs. 2.1 and 2.2. The electron and neutrino are emitted in 'opposite directions' if the relative helicity is positive, in the 'same direction' if the relative helicity is negative. This experiment thus measures the mean relative helicity $\langle h_{rel} \rangle$.

b. Pure Gamow-Teller transitions. The total angular momentum j_t of the lepton is one unit, and its projection m_t must be averaged over the three possible orientations. The results are given by Fig. 2.2. The electron and neutrino are emitted preferentially in the same direction if the relative helicity is positive, and in opposite directions if the relative helicity is negative. The effect is thus opposite to the Fermi case, and the magnitude of the effect is less by a factor $\frac{1}{3}$. This experiment thus also measures the mean relative helicity $\langle h_{rel} \rangle$.

c. Mixed transitions. There is no interference because no fixed axis is defined other than the direction of emission of the leptons. The resultant angular correlation is thus the sum of the Fermi and Gamow-Teller contributions, weighted by the relative magnitudes of the amplitudes. Because of the difference in sign and the factor $\frac{1}{3}$ in the Gamow-Teller case, the experiment measures the following linear combination of the mean helicities of the Fermi and Gamow-Teller components

$$\langle h_{rel} \rangle_F - (\tfrac{1}{3} \langle h_{rel} \rangle_{GT}) \frac{P_{GT}}{P_F},$$

where P_{GT}/P_F is the ratio of the Gamow-Teller transition probability to the Fermi transition probability.

3. *Experiments measuring the angular distribution of gamma*

TABLE 3.1
Quantities measured in beta decay experiments

Experiment		*Quantity measured*			
		In general	In Fermi transitions	In Gamow-Teller transitions	In mixed transitions
Electron polarization		$\langle h_e \rangle$	$\langle h_e \rangle$	$\langle h_e \rangle$	$\langle h_e \rangle$
Angular distribution from polarized nuclei	e	$\langle m_e h_e \rangle$	no effect	$\langle h_e \rangle$	$\langle f(h_e, a_F, a_{GT}) \rangle$
	ν	$\langle m_\nu h_\nu \rangle$	no effect	$\langle h_\nu \rangle$	$\langle f(h_\nu, a_F, a_{GT}) \rangle$
Circularly polarized correlation	β–γ	$\langle m_e h_e \rangle$	no effect	$\langle h_e \rangle$	$\langle f(h_e, a_F, a_{GT}) \rangle$
	ν–γ	$\langle m_\nu h_\nu \rangle$	no effect	$\langle h_\nu \rangle$	$\langle f(h_\nu, a_F, a_{GT}) \rangle$
e–ν angular correlation		$\langle m_e m_\nu h_e h_\nu \rangle$	$\langle h_e h_\nu \rangle$	$-\frac{1}{3}\langle h_e h_\nu \rangle$	$\langle h_e h_\nu \rangle_F - \frac{P_{GT}}{P_F} \cdot \frac{1}{3} \langle h_e h \rangle_{GT}$
Angular distribution of γ's from oriented nuclei		m_γ			$\left\lvert \frac{a_F}{a_{GT}} \right\rvert$

rays emitted after beta decay from polarized nuclei. The angular distribution of the gamma rays gives the intensity of the various angular momentum components in the outgoing electromagnetic wave. Since the initial nuclear polarization is known, the angular momentum carried by the leptons is determined by conservation of angular momentum as has been discussed in Chapter 2, § 5. This experiment is thus of interest in *mixed* allowed transitions, where the relative magnitudes of the Fermi and Gamow-Teller lepton amplitudes can be determined.

These results are conveniently summarized in Table 3.1.

We may now see how a complete set of experiments could be chosen to specify the lepton amplitudes in any particular allowed decay.

1. *Classification of the decay as allowed Fermi, Gamow-Teller, or Mixed* is achieved by the measurements of the spins and parities of the initial and final nuclear states.

2. *The average value of the electron helicity* $\langle h_e \rangle$ can be measured directly. In pure Gamow-Teller transitions it can also be measured by an angular distribution experiment with respect to a direction of nuclear or gamma ray polarization.

3. *The average value of the neutrino helicity* $\langle h_\nu \rangle$ can be measured by use of the conservation laws of momentum and angular momentum, if all other momenta, including that of the recoil nucleus are measured and all other relevant angular momenta are known. In pure Gamow-Teller transitions, it can also be measured by an angular distribution experiment with respect to a direction of nuclear or gamma ray polarization. In practice, it is of course the angular distribution of the recoil nucleus which is measured, not that of the neutrino.

4. *The average value of the relative helicity* $\langle h_{\text{rel}} \rangle$ can be measured in pure transitions by an electron-neutrino angular correlation experiment. In mixed transitions such an experiment gives a linear combination of the Fermi and Gamow-Teller components, with weighting factors of opposite sign whose magnitudes depend upon the magnitudes of the Fermi and Gamow-Teller amplitudes. It is of course the angular distri-

bution or energy distribution of the recoil which is measured, not of the neutrino.

5. *The relative magnitudes of the Fermi and Gamow-Teller amplitudes in a mixed transition* can be measured by an angular distribution experiment for gamma rays emitted after the beta decay from polarized or oriented nuclei. If the average relative helicities are known for both the Fermi and Gamow-Teller components, the mixing ratio is also given by an electron-neutrino angular correlation experiment. It is also obtainable from the angular distribution of either the electron or the neutrino with respect to a direction of nuclear or gamma ray polarization, provided that 1) the relative phase of the two amplitudes is known, 2) the helicity of the lepton is known and is not different for the Fermi and Gamow-Teller components.

6. *The relative phase of the Fermi and Gamow-Teller amplitudes* can be obtained from the angular distribution of either the electron or the neutrino with respect to a direction of nuclear or gamma ray polarization, provided that 1) the relative magnitudes of the two amplitudes is known, 2) the helicity of the lepton is known and is not different for the Fermi and Gamow-Teller components.

In practice, measurements of the nuclear spins and parities, the average electron helicity, and either the average neutrino helicity or the average relative helicity turn out to be sufficient to specify the lepton states completely in pure Gamow-Teller or pure Fermi transitions. This is because the neutrino always is found in an eigenstate of the helicity, either right-handed or left-handed. If this were not the case, two average helicities would not be sufficient to specify the lepton states because of the possibility of correlations; i.e. the average helicity of electrons emitted together with right-handed neutrinos might be different from the average helicity of electrons emitted with left-handed neutrinos.

In mixed transitions, the above measurements together with a measurement of the relative amplitudes and of the relative phase of the Fermi and Gamow-Teller components turn out to be sufficient to specify the lepton state completely. Here again the results happen to be of a particularly simple form making it unnecessary to look for complicated correlations which would be present in the general case.

The number of experiments required for complete specification of the states can also be understood by looking at the total number of independent amplitudes which must be determined. There are eight independent channels, each specified by a complex amplitude, thus giving sixteen real numbers to be determined. If we remove an overall phase factor and an overall normalization factor (we only are considering the relative magnitudes and phases of these amplitudes, not absolute values), we are left with fourteen numbers. A total of fourteen independent experiments would therefore be needed to specify the lepton states in the general case. The reason that fewer experiments suffice in practice is because many of the amplitudes vanish, and it is possible in a single experiment to show that several vanish.

In pure Gamow-Teller or Fermi transitions, the angular momentum selection rules tell us that only four of the eight channels are allowed (this is what we mean by pure transition). This reduces the number of independent measurements required from fourteen to six (two for each channel, less two for overall amplitude and phase factors), three amplitudes and three phases. The three amplitudes require three experiments, two average helicities and one helicity correlation, as was mentioned above. The three relative phases would involve measurements of transverse polarization and a correlated transverse polarization.

We are spared these complications because the results of the average helicity measurements seem to indicate that the neutrino is 100% polarized, thus eliminating two of the four channels corresponding to the opposite sign of neutrino polarization. The remaining two channels now require only two additional measurements. The average electron helicity gives the relative magnitudes of the two amplitudes. The relative phase would require a transverse polarization measurement, say of electrons

TABLE 3.2

The 'complete set' of experiments

8 Independent channels	14 experiments required	Angular momentum selection rules		Neutrino helicity measured as 100%	Electron helicity measured as 100%
		Pure F transitions	Pure GT transitions		
7 Relative amplitudes	Electron helicity F-component		X		100%
	Neutrino helicity F-component		X	100%	
	e–ν helicity correlation F-component		X	X	X
	Electron helicity GT-component	X			100%
	Neutrino helicity GT-component	X		100%	
	e–ν helicity correlation GT-component	X		X	X
	Relative magnitude F-GT components	X	X		
7 Relative phases	e–transverse polarization F-component		X		X
	ν–transverse polarization F-component		X	X	
	e–ν transv. pol. correlation F-component		X	X	X
	e–transverse polarization GT-component	X			X
	ν–transverse polarization GT-component	X		X	
	e–ν transv. pol. correlation GT-component	X		X	X
	Relative phase F-GT components	X	X		
Reduction in number of experiments		6 experiments remaining	6 experiments remaining	6 experiments remaining in mixed transitions; 2 in pure transitions	

Table 3.2 lists the complete set of experiments required to determine all the lepton amplitudes, and show how the number of experiments is reduced after the result of the measurements discussed above. An X indicates that the necessity of a particular experiment is eliminated as a result of a particular measurement; e.g. the measurement of neutrino helicity as 100% eliminates the need for a neutrino transverse polarization experiment.

emitted from polarized nuclei. However, at relativistic electron energies, $v_e \approx c$, the electrons are found to be 100% polarized, so that there is only one channel and no relative phase to be measured. At lower electron energies the polarization is less than 100%. However, although there is a phase factor to be determined, it is not of great interest, because beta decay theory predicts the energy dependence uniquely and the experiments are difficult.

In mixed transitions the situation is even more complicated in principle because all eight channels are present and the interference between the Gamow-Teller and Fermi components could well be different for each combination of helicities. However, the 100% polarization of neutrinos and relativistic electrons reduces the number of relevant channels to two, one Fermi and one Gamow-Teller. Thus only two additional experiments are required in addition to those for pure transition, to measure the relative magnitude and the relative phase.

CHAPTER 4

ANGULAR MOMENTUM IN BETA DECAY

4.1. GENERAL

A treatment of the '$1 + A \cos \theta$' experiments in the preceding chapters, based on the diagrams of Figs. 3.1 and 3.2, is over-simplified. Leptons are always pictured as going 'up' or 'down', always either parallel or antiparallel to one another. In reality there is always a continuous angular distribution, characterized by a '$1 + A \cos \theta$' distribution.

The reason for the discrepancy is that physicists are accustomed to tell lies when talking about angular momentum in quantum mechanics. They say *parallel* when they do not mean *parallel*, *antiparallel* when they do not mean *antiparallel*, and they say j^2 and j when they do not mean j^2 and j, they mean $j(j+1)$ and $\sqrt{j(j+1)}$. However, when they say m, they *really mean* m. Our diagrams tell the same kind of lies. To find out what these diagrams really mean, and to get an exact description of the angular distributions, we must go over our previous treatment and undo all the conventional lies.

When we say that a lepton has an angular momentum j, we really mean that the square of its angular momentum is $j(j+1)$. The magnitude of the angular momentum is therefore $\sqrt{j(j+1)}$. For the case which we call $j = \frac{1}{2}$, the angular momentum is really equal to $\sqrt{\frac{3}{2}}$, which is larger than $\frac{1}{2}$ by 73%. This is a big lie.

When we say that an angular momentum j is oriented 'parallel' to a fixed axis, we really mean that its projection on this axis, m, is equal to j. Since the magnitude of the angular momentum is not j, but $\sqrt{j(j+1)}$, the angular momentum is not oriented parallel to the axis, but rather at such an angle that its projection on the axis is equal to j. This is illustrated in Fig. 4.1. We see that a vector of magnitude $\sqrt{j(j+1)}$ having a projection j on the

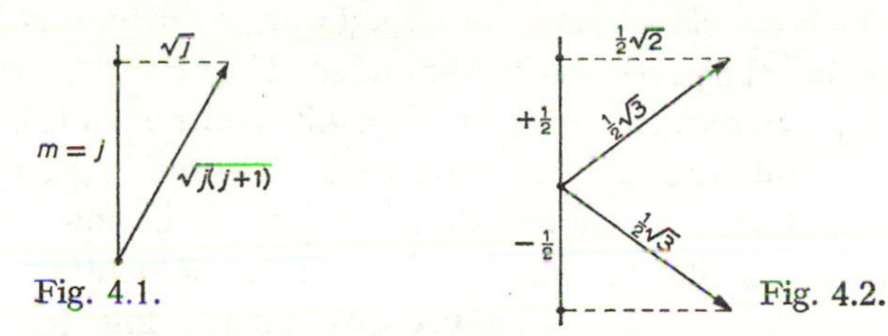

Fig. 4.1.

Fig. 4.2.

fixed axis has a component normal to the axis of $\sqrt{j}$. For the case of $j = \frac{1}{2}$, the component normal to the axis is $\frac{1}{2}\sqrt{2}$, which is 41 % *larger* than the component parallel to the axis, as shown in Fig. 4.2. Thus when we say that an angular momentum of $\frac{1}{2}$ is oriented '*parallel*' or '*antiparallel*' to an axis, we really mean '*nearly perpendicular*' in both cases, with a deviation from perpendicular of 35° either parallel or antiparallel to the axis. The direction of the component normal to the axis is not defined and it is not a constant of the motion; the vector precesses around the axis. The mean value of the normal component in any particular directions, (e.g. along the x-axis) is zero.

The same is true with regard to projections of angular momenta on directions other than fixed axes, such as directions of momenta. Thus when we talk of *right-handed* and *left-handed* leptons having their spins 'parallel' and 'antiparallel' to their momenta, we really mean precessing around their momenta at an angle of 55°. A right-handed electron in a state $m = \frac{1}{2}$ does not have both its spin and momentum pointing in the positive z-direction. Rather its spin is precessing around the z-axis and the momentum is precessing around its angular momentum. There is thus a continuous distribution for the angle between the momentum and the z-axis, as one would expect.

4.2. THE RELATION BETWEEN MOMENTUM, ANGULAR MOMENTUM AND HELICITY

Let us now consider a particle in a state of angular momentum j having a projection m on the z-axis. We should like to obtain the exact relation between the momentum, angular momentum and helicity for a particle in such a state, to replace the relation (3.1) which was based upon the oversimplified diagrams. Let us assume that the particle is *right*-handed. The orientations of the various vectors are illustrated in Fig. 4.3. The momentum of the particle is not always oriented upward, as in the oversimplified picture, but has a continuous distribution. Let us now calculate the mean value of the z-component p_z, of the momentum.

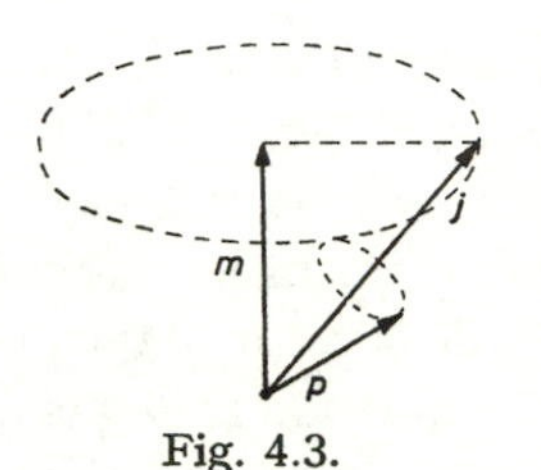

Fig. 4.3.

The vector diagram, Fig. 4.3, shows that the momentum vector $\boldsymbol{p}$ precesses around the angular momentum vector $\boldsymbol{j}$. There is no preferred direction defined normal to $\boldsymbol{j}$. Therefore the average value of the component of $\boldsymbol{p}$ which is normal to $\boldsymbol{j}$ must vanish. The only component of $\boldsymbol{p}$ whose average does not vanish is the component of $\boldsymbol{p}$ in the direction of the vector $\boldsymbol{j}$,

$$p_j = \frac{(\boldsymbol{p} \cdot \boldsymbol{j})\boldsymbol{j}}{j(j+1)}. \tag{4.1}$$

The mean value of the z-component of the momentum is thus

$$\langle p_z \rangle = \langle (p_j)_z \rangle = \frac{\langle (\boldsymbol{p} \cdot \boldsymbol{j}) j_z \rangle}{j(j+1)}. \tag{4.2}$$

Using equation (2.2) the expression $(\boldsymbol{p} \cdot \boldsymbol{j})$ can be expressed in terms of the helicity, h. In the state we have chosen, $j_z = m$. Thus

$$\langle p_z \rangle = \frac{p\langle hj_z \rangle}{2j(j+1)} = \frac{p\langle mh \rangle}{2j(j+1)}. \tag{4.3}$$

This result (4.3) is the true quantum-mechanical version of the relation (3.1) between momentum, angular momentum and helicity. However, we have proved it using a picture, Fig. 4.3, and we should be a bit suspicious of arguments based upon pictures at this point. Let us therefore now prove the results (4.2) and (4.3) rigorously, using the commutation rules of momenta and angular momenta.

Multiplying (4.2) by $j(j+1)$ and noting that this is the eigenvalue of the operator j^2 in any state $|j\rangle$ of angular momentum j, we obtain

$$\langle j|j^2 p_z - (\boldsymbol{j}\cdot\boldsymbol{p})j_z|j\rangle = 0\,. \tag{4.4}$$

Using a well-known vector identity, (4.4) can be written

$$\langle jm|\,[\boldsymbol{j}\times\{(\boldsymbol{p}\times\boldsymbol{j})-(\boldsymbol{j}\times\boldsymbol{p})\}_z]\,|jm\rangle = 0\,. \tag{4.5}$$

We can derive (4.5) and therefore also (4.2) using the commutation rules for momenta and angular momenta:

$$\begin{aligned} [j_x,j_y] &= \mathrm{i}j_z\,, & [j_x,p_y] &= [p_x,j_y] = \mathrm{i}p_z\,;\\ [j_y,j_z] &= \mathrm{i}j_x\,, & [j_y,p_z] &= [p_y,j_z] = \mathrm{i}p_x\,;\\ [j_z,j_x] &= \mathrm{i}j_y\,, & [j_z,p_x] &= [p_z,j_x] = \mathrm{i}p_y\,. \end{aligned} \tag{4.6}$$

The angular momentum operators are defined in units of $\hbar$, and have integral and half-integral eigenvalues. Using the commutation rules (4.6), we can calculate commutators involving the total angular momentum operator

$$j^2 = j_x^2 + j_y^2 + j_z^2\,, \tag{4.7}$$

such as

$$[j^2,\boldsymbol{p}] = \mathrm{i}\{(\boldsymbol{p}\times\boldsymbol{j})-(\boldsymbol{j}\times\boldsymbol{p})\} \tag{4.8a}$$

$$[j^2,(\boldsymbol{j}\times\boldsymbol{p})] = \mathrm{i}[\boldsymbol{j}\times\{(\boldsymbol{p}\times\boldsymbol{j})-(\boldsymbol{j}\times\boldsymbol{p})\}]\,. \tag{4.8b}$$

Let us now take the expectation values of (4.8) in the state $|j\rangle$. The left-hand sides of both equations have zero expectation values, since

$$\langle j|j^2\boldsymbol{p} - \boldsymbol{p}j^2|j\rangle = \{j(j+1) - j(j+1)\}\langle j|\boldsymbol{p}|j\rangle = 0,$$

and similarly for (4.8b). The expectation values of the right-hand sides must therefore vanish as well. Thus

$$\langle j|(\boldsymbol{p}\times\boldsymbol{j})-(\boldsymbol{j}\times\boldsymbol{p})|j\rangle = 0\,, \tag{4.9a}$$

$$\langle j|\boldsymbol{j}\times\{(\boldsymbol{p}\times\boldsymbol{j})-(\boldsymbol{j}\times\boldsymbol{p})\}|j\rangle = 0\,. \tag{4.9b}$$

Equations (4.9) state the results which we saw from the vector diagram; namely that the expectation values of those components of $\boldsymbol{p}$ which are normal to $\boldsymbol{j}$ must be zero. Furthermore, the z-component of (4.9b) is just (4.5). Since (4.5) is just an algebraic transformation of (4.2), we have now derived (4.5) and (4.2) in a rigorous manner.

The result (4.2) is a special case of a well-known general theorem that the expectation value of *any vector* in a state having a definite angular momentum j is given by the expectation value of the component of the vector which is parallel to j. The general result can be proved in a variety of ways, one of which is to use the same method which we have used to obtain (4.9), and the fact that the commutation rules (4.6) hold when $\boldsymbol{p}$ is replaced by *any vector*.

For the special case $j = \frac{1}{2}$, which is of interest in allowed transitions, (4.3) assumes the form

$$\begin{aligned} \langle p_z \rangle &= 2p \langle h j_z \rangle /3 = 2p \langle mh \rangle /3 \,, \\ \langle p_z \rangle /p &= \langle 2mh \rangle /3 \,. \end{aligned} \tag{4.10}$$

Equation (4.10) is the exact form of the relation (3.1) between momentum, angular momentum and helicity.

4.3. A FUNDAMENTAL RELATION FOR '$1+A\cos\theta$' EXPERIMENTS

The rigorous quantum-mechanical expressions for the relation between the mean values of momentum, angular momentum and helicity, (4.3) and (4.10), can be used to obtain an explicit expression for the parameter A appearing in the '$1 + A\cos\theta$' angular distributions.

We first note that the states of the leptons emitted in beta decay are outgoing spherical waves having momentum p. The asymmetry of the angular distribution with respect to the positive and negative z-axis can be described by the mean value of $\cos\theta$, the angle between the momentum vector of the lepton and the z-axis. Thus

$$\langle \cos\theta \rangle = \langle p_z/p \rangle = \frac{\langle hj_z \rangle}{2j(j+1)} = \frac{\langle mh \rangle}{2j(j+1)}. \tag{4.11}$$

If the angular distribution has the form '$1 + A\cos\theta$'

$$P(\theta)\mathrm{d}\Omega = \frac{1 + A\cos\theta}{4\pi}\,\mathrm{d}\Omega\,,$$

the mean value of $\cos\theta$ for such a distribution is

$$\langle \cos\theta \rangle = \int \cos\theta\, P(\theta)\mathrm{d}\Omega = \tfrac{1}{3}A\,. \tag{4.12}$$

The value of A can be obtained by equating (4.11) and (4.12). In general, the angular distribution of leptons emitted from beta decay would be more complicated than $1 + A\cos\theta$. The general distribution could be expressed as a power series in $\cos\theta$ or an expansion in Legendre polynomials. The maximum power of $\cos\theta$ which can occur in any given case is $2j$. Thus we see that for the case of allowed transitions, where $j = \frac{1}{2}$, the highest power of $\cos\theta$ which can occur is the first. There is therefore always a '$1 + A\cos\theta$' angular distribution in allowed transitions, and the value of A is given by

$$A = 2\langle hj_z \rangle = 2\langle mh \rangle\,. \tag{4.13}$$

Equation (4.13) is the rigorous expression of the conclusions drawn in Chapter 3, § 1 from the diagrams of Fig. 3.1. Note that equations (4.9) were obtained without requiring that the lepton state have a definite projection m of the total angular momentum on the z-axis. Equations (4.9) hold for any state which is an eigenfunction of j, regardless of whether it is also an eigenfunction of j_z. Equations (4.10) and (4.11) are therefore valid for any lepton state which is an eigenfunction of j and a mixture of the various m-states. They are also valid for the case where the leptons are not emitted in a pure state, but in an incoherent mixture of states all of the same value of j.

In the general case where the state is not an eigenfunction of j_z, the angular distribution can depend upon the polar angle ϕ as well as upon the angle θ. The results (4.11) and (4.12) still apply to this case. Equation (4.13) is valid for $j = \frac{1}{2}$, if the

'$1 + A \cos \theta$' distribution is interpreted as describing the number of particles emitted at an angle θ with respect to the z-axis, averaged over all polar angles ϕ.

4.4. OTHER RELEVANT PROPERTIES OF ANGULAR MOMENTUM

The results of all the '$1 + A \cos \theta$' experiments described in Chapter 3 can be obtained in a rigorous manner using (4.13) and arguments like those leading to (4.9). If the mean value of mh can be determined for the lepton in question, the angular distribution is given immediately.

There are some experiments, where the angular distribution is measured with respect to some other axis than the one initially chosen as the axis of quantization. For example, the leptons may be known to be in a definite m-state or mixture of m-states with respect to some axis such as the direction of polarization of a nucleus or gamma ray. Call this axis the z-axis. The question arises whether there will be any asymmetry in the angular distribution with respect to the x-axis or the y-axis. To answer this question we see from (4.13) that we need the mean value of j_x or j_y in a state which is an eigenstate of j_z or a known mixture of such eigenstates. That is, we need to know the matrix elements of j_x and j_y in the (j, m) representation. These are given in any elementary quantum mechanics text, such as SCHIFF [1955] (p. 145).

For allowed transitions, the leptons are in states of $j = \frac{1}{2}$. The results which we need are conveniently expressed in terms of the matrix element of the operators $j_x \pm ij_y$:

$$\langle j = \tfrac{1}{2}, m = +\tfrac{1}{2} | j_x + ij_y | j = \tfrac{1}{2}, m = -\tfrac{1}{2} \rangle = 1 \,, \quad (4.14a)$$

$$\langle j = \tfrac{1}{2}, m = -\tfrac{1}{2} | j_x - ij_y | j = \tfrac{1}{2}, m = +\tfrac{1}{2} \rangle = 1 \,. \quad (4.14b)$$

All other matrix elements of $j_x + ij_y$ between states of $j = \frac{1}{2}$ are zero.

4.5. LEPTON ANGULAR MOMENTA IN MIXED TRANSITIONS

In order to apply (4.13) to analyze '$1 + A \cos \theta$' experiments, it is necessary to know the mean value of the angular momentum m for a given lepton. This can be determined in pure transitions by considerations of angular momentum conservation, such as are discussed for simple cases in Chapter 2, § 7. The general cases are considered in Chapter 5. In mixed transitions, angular momentum conservation is not sufficient to determine the mean value of m, as was shown in Chapter 3, § 3. The mean value of m depends upon the relative magnitude and phase of the Fermi and Gamow-Teller amplitudes. Coherent contributions to $\langle m \rangle$ come from corresponding Fermi and Gamow-Teller channels having the same helicity quantum numbers (corresponding upper and lower diagrams in Fig. 3.1). We shall now show how the mean value of $\langle m \rangle$ can be determined in mixed transitions.

Let us consider a particular decay such as from a polarized nucleus, in which the leptons are emitted into a pure state having definite electron and neutrino helicities. Let a_{F} be the amplitude for emission into the Fermi channel, and let $\overset{+}{a}_{\mathrm{GT}}$, a_{GT}, and $\overset{-}{a}_{\mathrm{GT}}$ be the amplitudes for emission into the three Gamow-Teller channels, $m_{\mathrm{t}} = +1, 0$, and -1 respectively. Since the three Gamow-Teller amplitudes are related by angular momentum conservation, $\overset{+}{a}_{\mathrm{GT}}$ and $\overset{-}{a}_{\mathrm{GT}}$ can be expressed in terms of a_{GT} and the nuclear angular momenta. These relations are considered in the following chapter.

The wave function describing the leptons is a linear combination of the four basic states in the $(j_{\mathrm{t}}, m_{\mathrm{t}})$ representation. These are the Fermi state $(j_{\mathrm{t}} = 0, m_{\mathrm{t}} = 0)$ and the Gamow-Teller states $(j_{\mathrm{t}} = 1, m_{\mathrm{t}} = 0)$ and $(j_{\mathrm{t}} = 1, m_{\mathrm{t}} = \pm 1)$. For our purposes it is more convenient to express the wave function in terms of the basic set shown in Fig. 3.1; namely the eigenfunctions of m_{e} and m_{ν}. Let us use the following notation for these states:

$$\begin{aligned}
&|\mathrm{e}\uparrow \nu\uparrow\rangle \text{ is the state with } m_{\mathrm{e}} = +\tfrac{1}{2}, \quad m_{\nu} = +\tfrac{1}{2} \\
&|\mathrm{e}\downarrow \nu\uparrow\rangle \text{ is the state with } m_{\mathrm{e}} = -\tfrac{1}{2}, \quad m_{\nu} = +\tfrac{1}{2} \\
&|\mathrm{e}\uparrow \nu\downarrow\rangle \text{ is the state with } m_{\mathrm{e}} = +\tfrac{1}{2}, \quad m_{\nu} = -\tfrac{1}{2} \\
&|\mathrm{e}\downarrow \nu\downarrow\rangle \text{ is the state with } m_{\mathrm{e}} = -\tfrac{1}{2}, \quad m_{\nu} = -\tfrac{1}{2}
\end{aligned} \tag{4.15}$$

The Gamow-Teller and Fermi states having $m_{\mathrm{t}} = 0$ are respectively the symmetric (triplet) and antisymmetric (singlet) linear combinations of the two $m_t = 0$ states in the m_ν, m_e representation. Thus

$$|\mathrm{GT}, m_t = 0\rangle = \frac{1}{\sqrt{2}}(|e\uparrow \nu\downarrow\rangle + |e\downarrow \nu\uparrow\rangle)$$

$$|\mathrm{F}, m_t = 0\rangle = \frac{1}{\sqrt{2}}(|e\uparrow \nu\downarrow\rangle - |e\downarrow \nu\uparrow\rangle) \tag{4.16}$$

and the remaining Gamow-Teller states are

$$|\mathrm{GT}, m_t = 1\rangle = |e\uparrow \nu\uparrow\rangle$$

$$|\mathrm{GT}, m_t = -1\rangle = |e\downarrow \nu\downarrow\rangle .$$

Note that the phases used in defining these wave functions is entirely arbitrary. We could, for example, have obtained the opposite phases for the Fermi wave function by writing the neutrino first instead of the electron. The wave function describing the leptons is thus

$$|\mathrm{lept}\rangle = \overset{+}{a}_{\mathrm{GT}}|e\uparrow \nu\uparrow\rangle + \overset{-}{a}_{\mathrm{GT}}|e\downarrow \nu\downarrow\rangle + \frac{a_{\mathrm{GT}}}{\sqrt{2}}(|e\uparrow \nu\downarrow\rangle + |e\downarrow \nu\uparrow\rangle) + \frac{a_{\mathrm{F}}}{\sqrt{2}}(|e\uparrow \nu\downarrow\rangle - |e\downarrow \nu\uparrow\rangle)$$

$$= \overset{+}{a}_{\mathrm{GT}}|e\uparrow \nu\uparrow\rangle + \overset{-}{a}_{\mathrm{GT}}|e\downarrow \nu\downarrow\rangle + \frac{a_{\mathrm{GT}} + a_{\mathrm{F}}}{\sqrt{2}}|e\uparrow \nu\downarrow\rangle + \frac{a_{\mathrm{GT}} - a_{\mathrm{F}}}{\sqrt{2}}|e\downarrow \nu\uparrow\rangle . \tag{4.17}$$

Note that for a pure Fermi transition ($a_{\mathrm{GT}} = 0$) the amplitudes are symmetric with respect to $\nu\downarrow$ and $\nu\uparrow$, and also with respect to $e\downarrow$ and $e\uparrow$. No asymmetry would be observed in the lepton angular distributions in pure Fermi transitions. In pure Gamow-Teller transitions, the contribution of the $m_t = 0$ part (the last two terms), is also symmetric. Thus the contribution to the lepton angular distribution from the $m_t = 0$ part of the Gamow-Teller wave function is symmetric with respect to $+m$ and $-m$; all the asymmetry comes from the $m_t = \pm 1$ parts. In *mixed* transitions, however, the symmetry of the $m_t = 0$ part is destroyed and the $m_t = 0$ part contributes to the asymmetry (unless there is a phase shift of 90° between the Fermi and Gamow-Teller amplitudes).

The probability that the electron is emitted in a state of $m_{\mathrm{e}} = +\frac{1}{2}$ is obtained from (4.17) by taking the sum of the squares of the amplitudes corresponding to the states $|\mathrm{e}\uparrow\nu\uparrow\rangle$ and $|\mathrm{e}\uparrow\nu\downarrow\rangle$ and dividing by the normalization factor

$$N = |\overset{+}{a}_{\mathrm{GT}}|^2 + |\overset{-}{a}_{\mathrm{GT}}|^2 + |a_{\mathrm{GT}}|^2 + |a_{\mathrm{F}}|^2 . \tag{4.18}$$

There is no interference between $|\mathrm{e}\uparrow\nu\uparrow\rangle$ and $|\mathrm{e}\uparrow\nu\downarrow\rangle$ since no measurement is made on the neutrino and the neutrino states are orthogonal. Thus

$$P(\mathrm{e}\uparrow) = \frac{1}{N}\{|\overset{+}{a}_{\mathrm{GT}}|^2 + \tfrac{1}{2}|a_{\mathrm{GT}} + a_{\mathrm{F}}|^2\} .$$

Similarly,

$$\begin{aligned} P(\mathrm{e}\downarrow) &= \frac{1}{N}\{|\overset{-}{a}_{\mathrm{GT}}|^2 + \tfrac{1}{2}|a_{\mathrm{GT}} - a_{\mathrm{F}}|^2\} , \\ P(\nu\uparrow) &= \frac{1}{N}\{|\overset{+}{a}_{\mathrm{GT}}|^2 + \tfrac{1}{2}|a_{\mathrm{GT}} - a_{\mathrm{F}}|^2\} , \\ P(\nu\downarrow) &= \frac{1}{N}\{|\overset{-}{a}_{\mathrm{GT}}|^2 + \tfrac{1}{2}|a_{\mathrm{GT}} + a_{\mathrm{F}}|^2\} . \end{aligned} \tag{4.19}$$

Note that $P(\mathrm{e}\uparrow) + P(\mathrm{e}\downarrow) = P(\nu\uparrow) + P(\nu\downarrow) = 1$, as expected. The mean values of m_{e} and m_ν are thus:

$$\langle m_{\mathrm{e}}\rangle = \tfrac{1}{2}[P(\mathrm{e}\uparrow) - P(\mathrm{e}\downarrow)] = \frac{1}{N}\{|\overset{+}{a}_{\mathrm{GT}}|^2 - |\overset{-}{a}_{\mathrm{GT}}|^2 + 2|a_{\mathrm{GT}}a_{\mathrm{F}}|\cos\phi\} \tag{4.20a}$$

$$\langle m_\nu\rangle = \tfrac{1}{2}[P(\nu\uparrow) - P(\nu\downarrow)] = \frac{1}{N}\{|\overset{+}{a}_{\mathrm{GT}}|^2 - |\overset{-}{a}_{\mathrm{GT}}|^2 - 2|a_{\mathrm{GT}}a_{\mathrm{F}}|\cos\phi\} , \tag{4.20b}$$

where ϕ is the phase angle between the Fermi and Gamow-Teller amplitudes,

$$\mathrm{e}^{\mathrm{i}\phi} = \frac{a_{\mathrm{F}}/a_{\mathrm{GT}}}{|a_{\mathrm{F}}/a_{\mathrm{GT}}|} . \tag{4.21}$$

The first two terms in each expression (4.20) represent the effect in pure Gamow-Teller transitions. The third term is the interference contribution. Note that the interference has opposite signs for the electron and neutrino distributions. Thus, if the asymmetry for one of the leptons is *greater* than would be the case for a pure Gamow-Teller transition, the asymmetry for the other lepton must be *less* than for the pure Gamow-Teller transition. This is in accord with the simple argument of Chapter 3, § 3. The interference comes from the $m_t = 0$ component, where the sum of the mean values of m_e and m_ν must be equal and opposite. On the other hand the asymmetry in the lepton angular distributions in pure Gamow-Teller transitions comes from the $m_t = 1$ part, where m_e and m_ν have the same sign.

CHAPTER 5

ALLOWED BETA DECAY WITH FORMULAS

We can now revise our calculations of the '$1 + A\cos\theta$' experiments in Chapter 3, using the exact expressions (4.10) and (4.13) for the relation between momentum, angular momentum and helicity instead of the oversimplified expressions (3.1) and (3.2).

5.1. ELECTRON-NEUTRINO ANGULAR CORRELATION

Let us take our axis of quantization in the direction of emission of the neutrino. Then the projection of the neutrino angular momentum m_ν on this axis is simply

$$m_\nu = \tfrac{1}{2}h_\nu\,, \tag{5.1}$$

since the helicity is the sign of the projection of the angular momentum on the same axis. The angular distribution of the electron with respect to this axis is '$1 + A\cos\theta$', with the asymmetry parameter A given by (4.13) in terms of the electron helicity h_e and the projection of the electron angular momentum m_e on the axis of quantization. In order to express this in terms of the *relative helicity*, as in the simple treatment, we need to express m_e in terms of the neutrino helicity, using conservation of the angular momentum and (5.1).

A. *Pure Fermi transitions.* In this case $m_t = 0$, and $m_e = -m_\nu = -\frac{1}{2}h_\nu$. Thus

$$A = -\langle h_e h_\nu\rangle = -\langle h_{\mathrm{rel}}\rangle\,. \tag{5.2}$$

B. *Pure Gamow-Teller transitions.* Here $m_t = 1, 0,$ or -1. If no nuclear polarization measurements or selections are made, and if the axis of quantization is arbitrarily chosen, the three values of m_t occur with equal probability. If

$$m_t = \pm 1, \qquad m_e = m_\nu ;$$

if

$$m_t = 0, \qquad m_e = -m_\nu .$$

Thus

$$\langle m_e \rangle = \tfrac{1}{3} \langle m_\nu \rangle . \tag{5.3}$$

In the particular case under consideration, the axis of quantization is not chosen arbitrarily, but in the direction of emission of the neutrino. Thus all values of m_t do not occur with equal probability. For example, if the neutrino is right-handed, then $m_\nu = +\frac{1}{2}$ and the case $m_t = -1$ must be excluded. However, the relation (5.3) is still valid. This can be seen as follows:

In selecting those decays in which neutrinos are emitted in the direction of our axis, we are selecting decays in which $m_\nu = \frac{1}{2} h_\nu$ as given by (5.1). Let us first consider the case where neutrinos are *right-handed*; then we are selecting states of $m_\nu = +\frac{1}{2}$. We are therefore rejecting states of $m_t = -1$. The state $m_t = 0$ is an equal mixture of the states $m_\nu = +\frac{1}{2}$ and $m_\nu = -\frac{1}{2}$, while the state $m_t = +1$ consists entirely of $m_\nu = +\frac{1}{2}$. Therefore the probability of selection of an $m_t = +1$ state is twice the probability of selection of an $m_t = 0$ state. Thus (5.3) is also valid for this case. A similar analysis shows that (5.3) is also valid for left-handed neutrinos and for an arbitrary helicity mixture.

We thus have

$$A = \tfrac{1}{3} \langle h_e h_\nu \rangle = \tfrac{1}{3} \langle h_{rel} \rangle . \tag{5.4}$$

C. *Mixed transitions*. As has been discussed in Chapter 3, § 3, there is no interference between the Fermi and Gamow-Teller components in this experiment. The asymmetry parameter is given by combining (5.2) and (5.4) with weighting factors appropriate to the particular decay. It is customary to express the degree of mixing in terms of the Fermi fraction x defined as the relative probability that the leptons emitted are in a $j_t = 0$ state.

The probability that they are in a $j_t = 1$ state (the Gamow-

Teller fraction) is then $1-x$. The angular distribution asymmetry parameter

$$A = -x\langle h_{\text{rel}}\rangle_{\text{F}} + \tfrac{1}{3}(1-x)\langle h_{\text{rel}}\rangle_{\text{GT}} \tag{5.5}$$

where $\langle h_{\text{rel}}\rangle_{\text{F}}$ and $\langle h_{\text{rel}}\rangle_{\text{GT}}$ are the mean relative helicities of the Fermi and Gamow-Teller components respectively.

The results of the electron-neutrino angular correlation experiment are conveniently displayed in a Scott diagram (KONOPINSKI [1959]) as shown in Fig. 5.1, plotting A against x. Each measurement is represented by a point on the diagram. If the mean relative helicities $\langle h_{\text{rel}}\rangle_{\text{F}}$ and $\langle h_{\text{rel}}\rangle_{\text{GT}}$ are assumed to be constants characteristic of beta decay which do not vary from one nucleus to another, these points should all be on a straight line. The lines for the four limiting cases $\langle h_{\text{rel}}\rangle_{\text{F}} = \pm 1$ and $\langle h_{\text{rel}}\rangle_{\text{GT}} = \pm 1$ are shown in Fig. 5.1. These four cases have a

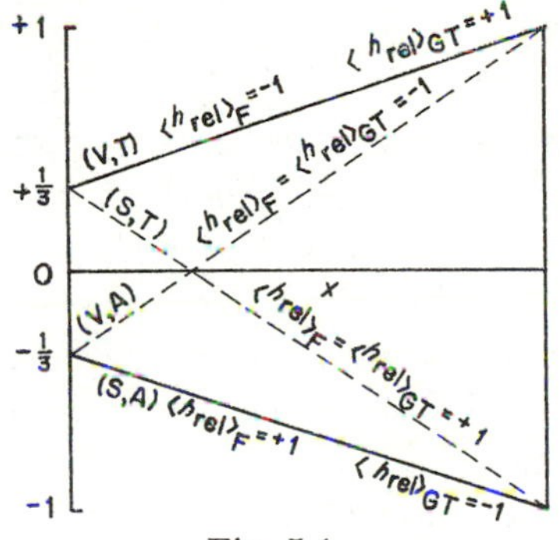

Fig. 5.1.

simple interpretation in beta decay theory, as will be discussed in Chapter 6, Fig. 6.1 corresponding to the presence of only two of the four beta decay interactions (SVTA). Each line is labeled by the appropriate pair of interactions.

5.2. ANGULAR DISTRIBUTION OF LEPTONS FROM POLARIZED NUCLEI

Let us choose our axis of quantization in the direction of polarization of the nucleus. Thus, if the initial nuclear state has spin

J_i, the projection of the spin on our axis is $M_i = J_i$. The spin of the final nuclear state J_f can be either $J_i - 1$, J_i, or $J_i + 1$ in allowed transitions. The sum of the vector $\boldsymbol{J}_f$ and the total lepton angular momentum $\boldsymbol{j}_t$ must be equal to the vector $\boldsymbol{J}_i$, as shown in Fig. 5.2.

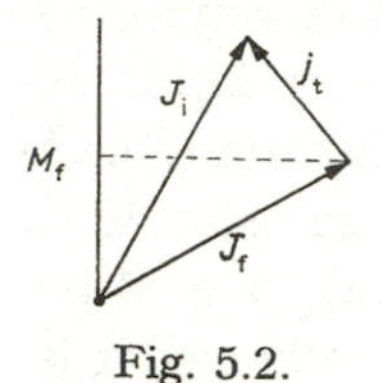

Fig. 5.2.

The angular distribution of the lepton with respect to the axis is $1 + A \cos \theta$, with A given by (4.13). We need only to determine the mean value of m in terms of the other angular momenta to obtain the result. In the simple case considered in Fig. 2.2, $J_f = J_i - 1$, m is determined uniquely and is easily calculated. We now consider the general case.

A. *Pure Fermi transitions*. There is no effect, $A = 0$, as was pointed out in the simple treatment of Chapter 3, § 4. Since $j_t = 0$, the lepton state must be spherically symmetric and

$$\langle m_e \rangle = \langle m_\nu \rangle = 0 .$$

B. *Pure Gamow-Teller transitions*. The mean values of m_e and m_ν can be obtained from the vector diagram Fig. 5.2 and arguments similar to those used in Chapter 4 to derive (4.2).

The vector $\boldsymbol{J}_i$ remains at a fixed angle with respect to the axis, such that its projection $M_i = J_i$. The triangle formed by the vectors $\boldsymbol{J}_i$, $\boldsymbol{J}_f$, and $\boldsymbol{j}_t$ precesses around $\boldsymbol{J}_i$. There is no preferred direction normal to $\boldsymbol{J}_i$; thus the component of $\boldsymbol{j}_t$ normal to $\boldsymbol{J}_i$ is zero. The mean value of the z-component of $\boldsymbol{j}_t$ is thus equal to the mean value of the z-component of the projection of $\boldsymbol{j}_t$ on $\boldsymbol{J}_i$,

$$\langle m_t \rangle = \langle (\boldsymbol{j}_t)_z \rangle = \frac{\langle (\boldsymbol{j}_t \cdot \boldsymbol{J}_i) M_i \rangle}{J_i(J_i + 1)} . \tag{5.6}$$

Equation (5.6) is directly analogous to (4.2). A formal proof using commutation relations can be also given, following exactly the lines of (4.9). The total angular momentum of the final state *of the whole system*, $\boldsymbol{J}_F = \boldsymbol{j}_t + \boldsymbol{J}_f$, is analogous to the angular momentum $\boldsymbol{j}$ in our previous example, and the vector $\boldsymbol{j}_t$ is analogous to the vector $\boldsymbol{p}$. Since $\boldsymbol{J}_F$ and its projection on the axis M_F must be equal respectively to $\boldsymbol{J}_i$ and M_i by angular momentum conservation, (5.6) follows from exactly the same manipulations as in (4.9), using commutators of $\boldsymbol{J}_F$ and $\boldsymbol{j}_t$.

The vectors $\boldsymbol{j}_e$ and $\boldsymbol{j}_\nu$ precess around their sum $\boldsymbol{j}_t$ in exactly the same fashion, and the component of $\boldsymbol{j}_e$ normal to $\boldsymbol{j}_t$ also averages to zero. Since $\boldsymbol{j}_e$ and $\boldsymbol{j}_\nu$ are vectors of equal magnitude, the projections of $\boldsymbol{j}_e$ and $\boldsymbol{j}_\nu$ on $\boldsymbol{j}_t$ are each $\frac{1}{2}\boldsymbol{j}_t$. We therefore obtain from (4.13) and (5.6)

$$A_l = 2\langle m_l h_l \rangle = \langle m_t h_l \rangle = \frac{\langle h_l (\boldsymbol{j}_t \cdot \boldsymbol{J}_i) M_i \rangle}{J_i (J_i + 1)} . \tag{5.7}$$

In (5.7) the subscript l denotes either e or ν as the same expression holds for both the electron and neutrino distributions. The quantity $(\boldsymbol{j}_t \cdot \boldsymbol{J}_i)$ can be expressed using the law of cosines

$$2\boldsymbol{j}_t \cdot \boldsymbol{J}_i = j_t(j_t + 1) + J_i(J_i + 1) - J_f(J_f + 1) . \tag{5.8}$$

Substituting (5.8) into (5.7) and noting that $M_i = J_i$ and $j_t = 1$,

$$A_l = \frac{\langle h_l \rangle \, [2 + (J_i - J_f)(J_i + J_f + 1)]}{2(J_i + 1)} . \tag{5.9}$$

Substituting the three possible values for $J_i - J_f$ into (5.9) we obtain

$$A_l = \langle h_l \rangle \qquad (J_i = J_f + 1) \tag{5.10a}$$

$$A_l = \frac{\langle h_l \rangle}{(J_i + 1)} \qquad (J_i = J_f) \tag{5.10b}$$

$$A_l = -\frac{\langle h_l \rangle J_i}{(J_i + 1)} \qquad (J_i = J_f - 1) . \tag{5.10c}$$

Note that (5.10a) is in agreement with the result of the simple arguments used with Fig. 2.2. Equation (5.10b) and (5.10c) show that the asymmetry is reduced when both values for m are possible, as is to be expected.

C. *Mixed transitions*. The mean values of m_e and m_ν are given by (4.20). Substituting into (4.13) we obtain the asymmetry parameters

$$A_e = \frac{1}{N} \langle h_e \rangle \left[|\overset{+}{a}_{GT}|^2 - |\overset{-}{a}_{GT}|^2 + 2|a_{GT} a_F| \cos\phi \right] \tag{5.11a}$$

$$A_\nu = \frac{1}{N} \langle h_\nu \rangle \left[|\overset{+}{a}_{GT}|^2 - |\overset{-}{a}_{GT}|^2 - 2|a_{GT} a_F| \cos\phi \right] . \tag{5.11b}$$

where N is given by equation (4.18).

Expressions for a^+_{GT} and a^-_{GT} in the case $M_i = +J_i$ are obtained using (5.10b). Since M_f can only be J_i and $J_i - 1$, m_t can only be 0 or $+1$. Thus a^-_{GT} must vanish. Equation (5.10b) tells us that the mean value of m_t in the Gamow-Teller contribution to a $J_i = J_f$ transition is

$$\langle m_t \rangle_{GT} = 2 \langle m_e \rangle_{GT} = \frac{1}{(J_i + 1)} .$$

Since the $m_t = 0$ channel does not contribute to $\langle m_t \rangle$, the probability that $m_t = 1$ must be $1/(J_i + 1)$. Then the probability that $j_t = 1$, $m_t = 0$ must be the remaining $J_i/(J_i + 1)$. This gives us

$$\overset{-}{a}_{GT} = 0 , \tag{5.12a}$$

$$|\overset{+}{a}_{GT}|^2 = \frac{|a_{GT}|^2}{J_i} . \tag{5.12b}$$

The asymmetry parameter A is obtained by substituting (5.12) into (5.11).

$$A_e = \langle h_e \rangle \frac{|a_{GT}^2| + 2J_i |a_F a_{GT}| \cos\phi}{(J_i + 1)|a_{GT}^2| + J_i |a_F^2|} \tag{5.13a}$$

$$A_\nu = \langle h_\nu \rangle \frac{|a_{GT}^2| - 2J_i |a_F a_{GT}| \cos\phi}{(J_i + 1)|a_{GT}^2| + J_i |a_F^2|} . \tag{5.13b}$$

For the case $J_i = \frac{1}{2}$ which includes the neutron decay

$$A_e = \langle h_e \rangle \frac{2|a_{GT}|^2 + 2|a_{GT} a_F| \cos\phi}{3|a_{GT}|^2 + |a_F|^2} , \tag{5.14a}$$

$$A_\nu = \langle h_\nu \rangle \frac{2|a_{\mathrm{GT}}|^2 - 2|a_{\mathrm{GT}}a_{\mathrm{F}}| \cos \phi}{3|a_{\mathrm{GT}}|^2 + |a_{\mathrm{F}}|^2}. \tag{5.14b}$$

Here we see again that the neutrino distribution depends upon the Fermi and Gamow-Teller amplitudes in exactly the same way as the electron distribution, but the dependence upon the relative phase is exactly opposite. The phase $\phi = 0°$ gives maximum electron asymmetry; the phase $\phi = 180°$ gives maximum neutrino asymmetry and minimum electron asymmetry. Since the choice of phase factors in defining our wave functions was arbitrary, it is a matter of convention which phase is called 0° and which is called 180°. However, it is clear that one phase gives maximum electron asymmetry and minimum neutrino asymmetry and that the opposite phase gives the reverse.

Measurements of both the electron and neutrino angular distributions gives both the relative amplitudes and the relative phases of the Fermi and Gamow-Teller components. However, if ϕ is near 0° or 180°, as is the case in the neutron decay, the result is not sensitive to small changes in ϕ. It is of interest to know whether ϕ is exactly 0° or 180°; i.e. whether the amplitudes can be both taken as *real*, for reasons associated with time reversal invariance. For this, a different experiment has been used to determine an upper limit for the possible value of the phase angle. This experiment involves measurement of the electron-neutrino angular correlation in the plane perpendicular to the direction of polarization of the nucleus. This experiment is discussed in section 5 of this chapter.

With $\phi = 180°$ in the neutron decay, as it seems to be, the measurement of the electron angular distribution gives the ratio of the Fermi and Gamow-Teller amplitudes. From (5.14a) we note that the asymmetry is proportional to the *difference* between the two amplitudes and vanishes if the two amplitudes are equal. Such an experiment is therefore a sensitive way to measure the relative magnitudes of the Fermi and Gamow-Teller amplitudes.

In decays from complex nuclei, the relative magnitudes of the

Fermi and Gamow-Teller contributions to a given decay depends upon the structure of the nuclei and are most conveniently expressed in terms of the Fermi fraction x defined in the previous section

$$x = \frac{|a_F|^2}{|a_F|^2 + |a_{GT}|^2 + |a_{GT}^+|^2 + |a_{GT}^-|^2}. \tag{5.15a}$$

Using (5.12)

$$x = \frac{|a_F|^2}{|a_F|^2 + |a_{GT}|^2 (J_i + 1)/J_i}. \tag{5.15b}$$

Then

$$\frac{|a_{GT}|^2}{|a_F|^2} = \frac{1-x}{x} \frac{J_i}{J_i + 1}. \tag{5.15c}$$

Substituting (5.12) and (5.15) into (5.11) we obtain

$$\begin{aligned} A_e &= \langle h_e \rangle [(1-x) + 2\sqrt{x(1-x) J_i(J_i+1)} \cos\phi]/(J_i+1) \\ A_\nu &= \langle h_\nu \rangle [(1-x) - 2\sqrt{x(1-x) J_i(J_i+1)} \cos\phi]/(J_i+1) \end{aligned} \tag{5.16}$$

The first term in the square bracket in (5.16) is the pure Gamow-Teller term; the second is the interference term. Note that (5.16) reduces to (5.10b) for $x = 0$. As J_i increases, the contribution of the Gamow-Teller term in (5.16) decreases and the interference term increases. For the case of $J_i = J_f = \frac{1}{2}$ the two terms are equal at $a_{GT} = a_F$ and the interference term is never greater than the Gamow-Teller term. The maximum destructive interference can only reduce the asymmetry to zero. For $J_i > \frac{1}{2}$, the interference term can become greater than the Gamow-Teller term and reverse the sign of the asymmetry. For large J_i, the interference term predominates. This effect can be understood by examining Fig. 5.2 and remembering the usual angular momentum lies. For large J_i, $J_i(J_i + 1)$ is very nearly equal to J_i^2, and the nuclear polarization vector becomes very nearly *really* parallel to the z-axis. The vector $\boldsymbol{j}_t$ is very small in comparison with $\boldsymbol{J}_i$ and must therefore be very nearly perpendicular to the z-axis since $J_f = J_i$. Thus m_t is very nearly equal to zero; i.e. the probability that it is -1 is very small.

5.3. ANGULAR DISTRIBUTION OF LEPTONS FROM AND TO PARTIALLY POLARIZED NUCLEI

Let us now generalize the results of the preceding section to apply to the following situations which occur frequently in experiments:

a. The nuclei are not 100% polarized in the initial state but are only partially polarized;

b. The nuclei are unpolarized in the initial state, but the final state nuclei are selected to have a net partial polarization. (This is generally done by observing a subsequently emitted polarized gamma ray.)

The case of partial polarization can be treated easily by the use of equation (5.7), which gives the asymmetry parameter A for the lepton angular distribution. No specific assumptions about the initial state of polarization have yet been used explicitly. The result for the general case of partial polarization is obtained by keeping M_i as an independent parameter, rather than setting $M_i = J_i$ as was done in (5.9). The results obtained in this way, analogous to (5.9), (5.10) and (5.16) have the additional factor $\langle M_i \rangle / J_i$, which is just the degree of initial polarization.

$$A_l = \frac{\langle M_i h_l \rangle \, [2 + (J_i - J_f)(J_i + J_f + 1)]}{2J_i(J_i + 1)} \tag{5.9'}$$

$$A_l = \frac{\langle M_i h_l \rangle}{J_i} \qquad (J_i = J_f + 1) \tag{5.10'a}$$

$$A_l = \frac{\langle M_i h_l \rangle}{J_i(J_i + 1)} \qquad (J_i = J_f) \tag{5.10'b}$$

$$A_l = -\frac{\langle M_i h_l \rangle}{J_i + 1} \qquad (J_i = J_f - 1) \tag{5.10'c}$$

$$A_l = \frac{\langle M_i h_l \rangle \, [(1 - x) \pm 2\sqrt{x(1 - x) J_i(J_i + 1)} \cos \phi]}{J_i(J_i + 1)}. \tag{5.16'}$$

(+ sign for e; − sign for ν)

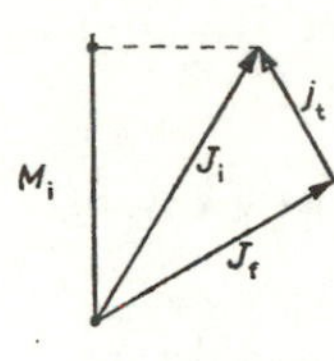

Fig. 5.3.

The vector diagram of Fig. 5.3 represents the case where the nucleus is initially unpolarized, but the final state is selected to be polarized. We assume that only nuclei in a definite final state M_f are selected. Again the vector $\boldsymbol{J}_i$ is the vector sum $\boldsymbol{J}_f + \boldsymbol{j}_t$. However, here the vector $\boldsymbol{J}_f$ remains at a fixed angle with respect to the z-axis; namely that which makes the projection of $\boldsymbol{J}_f$ on the axis equal to M_f. The orientation of the plane of the triangle ($\boldsymbol{J}_i, \boldsymbol{j}_t, \boldsymbol{J}_f$) is arbitrary; there is no preferred direction defined normal to $\boldsymbol{J}_f$. Thus the component of $\boldsymbol{j}_t$ normal to $\boldsymbol{J}_f$ averages to zero, and the mean value of the z-component of $\boldsymbol{j}_t$ is obtained by projection on $\boldsymbol{J}_f$,

$$\langle m_t \rangle = \langle (\boldsymbol{j}_t)_z \rangle = \frac{\langle (\boldsymbol{j}_t \cdot \boldsymbol{J}_f) M_f \rangle}{J_f(J_f + 1)}. \qquad (5.17)$$

Using the law of cosines and substituting into (4.13) we obtain the asymmetry parameter

$$A_l = \frac{\langle M_f h_l \rangle [(J_i - J_f)(J_i + J_f + 1) - 2]}{2J_f(J_f + 1)}. \qquad (5.18)$$

Note that (5.18) can be obtained directly from (5.9′) by interchanging J_i and J_f and reversing the sign. This is to be expected, since the roles of J_i and J_f have been interchanged in the vector diagram. The sign is reversed because $j_t = J_i - J_f$ changes sign when J_i and J_f are interchanged. Thus the results of (5.10′) and (5.16′) can be applied to this case simply by interchanging J_i and J_f and reversing the sign of A.

$$\begin{aligned} A_l &= -\frac{\langle M_f h_l \rangle}{J_f} = -\frac{\langle M_f h_l \rangle}{(J_i + 1)} && (J_f = J_i + 1) \\ A_l &= -\frac{\langle M_f h_l \rangle}{J_i(J_i + 1)} && (J_f = J_i) \qquad (5.10'') \\ A_l &= +\frac{\langle M_f h_l \rangle}{(J_f + 1)} = \frac{\langle M_f h_l \rangle}{J_i} && (J_f = J_i - 1) \end{aligned}$$

and

$$A = -\frac{\langle M_f h_i \rangle [(1-x) \pm 2\sqrt{x(1-x)J_f(J_f+1)}\cos\phi]}{J_f(J_f+1)} \tag{5.16''}$$

(+ sign for e, — sign for ν)

5.4. β–γ CIRCULARLY POLARIZED CORRELATION

We have seen in the simplified treatment of Chapters 1–3 that this type of experiment is equivalent to the angular distribution from polarized nuclei. In both experiments the mean value of the lepton angular momentum is determined by the use of angular momentum conservation. The analysis of this case is similar to that of the preceding case, except that there are more angular momentum vectors and the diagram is more complicated.

Let us consider the case where a nucleus having angular momentum J_i decays by beta decay to a state of angular momentum J_f, and then by emission of a pure 2^L multipole gamma ray to a state of angular momentum J' as shown in Fig. 5.4. The cir-

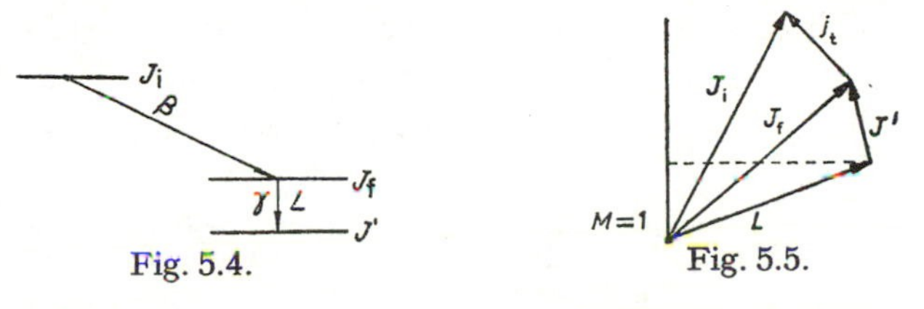

Fig. 5.4. Fig. 5.5.

cular polarization of the gamma ray is measured. Let us take the direction of emission of the gamma ray as our axis of quantization. The angular momenta are represented in the diagram of Fig. 5.5. The gamma ray momentum $\boldsymbol{L}$ has a projection M on the axis of quantization. M must be $+1$ or -1 depending upon whether the gamma ray has right-handed or left-handed helicity. The sum of the vectors $\boldsymbol{L}$ and $\boldsymbol{J}'$ is equal to the vector $\boldsymbol{J}_f$ and the triangle $(\boldsymbol{L}\,\boldsymbol{J}'\boldsymbol{J}_f)$ precesses around the vector $\boldsymbol{J}_f$ which remains at a fixed angle with respect to the vector $\boldsymbol{L}$.

From the vector diagram, we see that this case is a combination of the two cases treated in the preceding section. The leptons are emitted from unpolarized nuclei, but the final state J_f of the beta transition is selected to have a net partial polarization. We can calculate the value of this partial polarization directly from the vector diagram.

The vector $\boldsymbol{J}_f$ precesses around $\boldsymbol{L}$, and only the component of $\boldsymbol{J}_f$ in the direction of $\boldsymbol{L}$ has a non-vanishing mean value. Thus

$$\langle M_f \rangle = \langle J_f \rangle_z = \frac{\langle (\boldsymbol{J}_f \cdot \boldsymbol{L}) M \rangle}{L(L+1)}$$

$$= \frac{M[L(L+1) + (J_f - J')(J_f + J' + 1)]}{2L(L+1)} . \tag{5.19}$$

The value of the asymmetry parameter A for the various types of transitions are then obtained by substituting (5.19) into (5.18), (5.10″), and (5.16″).

The result (5.19) can be expressed more simply if a particular value of L is chosen. The most useful values are $L = \pm (J' - J_f)$, corresponding to the *lowest* multipole which can be emitted with conservation of angular momentum. The result (5.19) is valid only for a *pure* multipole of order L. If L is not the lowest multipole which can be emitted, the lower multipoles will also be emitted and the transition will not be pure. The only exception to this case is $J' - J_f = 0$, as there are no monopole gamma rays. We therefore write

$$\begin{aligned} \langle M_f \rangle &= \frac{(J_f + 1)M}{(L+1)} & J' &= J_f - L \\ \langle M_f \rangle &= \frac{-J_f M}{(L+1)} & J' &= J_f + L \\ \langle M_f \rangle &= \tfrac{1}{2} M & J' &= J_f . \end{aligned} \tag{5.20}$$

Since the projection of the angular momentum of the gamma ray, M, on its direction of emission can only be ± 1, we see that the degree of partial polarization of the final state, $\langle M_f \rangle / J_f$, is approximately $1/(L+1)$ and therefore appreciably less than unity, unless L and J_f are both small.

If the gamma ray is not a pure multipole, but is a mixture of several L values, (5.19) can be used for each component individually. Since the operators J_{z_f} and L commute, there are no cross terms in the mean value of J_{z_f} from different values of L.

Equation (5.19) can be rewritten for this case in the form

$$\langle M_f \rangle = \frac{M}{2}\left[1 + (J_f - J')(J_f + J' + 1)\left\langle\frac{1}{L(L+1)}\right\rangle\right], \qquad (5.19')$$

where $\langle 1/L(L+1)\rangle$ is the mean value of $(1/L(L+1))$ for the particular gamma ray transition.

The asymmetry parameter A for the lepton angular distribution can now be obtained for the three different types of allowed transitions.

As has been pointed out in Chapter 1, § 3 the asymmetry parameter A is not measured directly in practical cases. Rather than measure an angular distribution it is more convenient to reverse the direction of polarization of the gamma ray polarization analyzer and to measure at a fixed angle. Thus measurements are made, one for $M = +1$ and one for $M = -1$. Since the asymmetry parameter A is proportional to M, reversing M also reverses A, i.e. rotates the $1 + A\cos\theta$ angular distribution by 180°. Thus measurement at one angle and at $M = +1$ and -1 is equivalent to measurement for one value of M at two angles differing by 180°.

A. *Pure Fermi transitions.* As in the angular distribution from polarized nuclei, there is no effect in pure Fermi transitions, where $j_t = m_t = 0$.

B. *Pure Gamow-Teller transitions.* The asymmetry parameter A is obtained directly by substituting (5.19) into (5.10″). The results have the form

$$A_l = \langle h_l M\rangle f(J_i, J_f, J', L), \qquad (5.21)$$

where

$$f(J_i, J_f, J', L) = \frac{[(J_i - J_f)(J_i + J_f - 1) - 2][L(L+1) + (J_f - J')(J_f + J' + 1)]}{2J_i(J_i + 1)\; 2L(L+1)}$$

for a pure gamma ray multipole, or

$$= \frac{[(J_i - J_f)(J_i + J_f - 1) - 2]}{2J_i(J_i + 1)} \times \tfrac{1}{2}\left[1 + (J_f - J')(J_f + J' + 1)\left\langle\frac{1}{L(L+1)}\right\rangle\right]$$

for mixed multipoles.

The values of f are tabulated in Table 5.1 for the most important values of L.

TABLE 5.1

Values of $f(J_i, J_f, J', L)$

	$J' = J_f - L$	$J' = J_f$	$J' = J_f + L$
$J_f = J_i - 1$	$\frac{1}{(L+1)}$	$\frac{1}{2J_i}$	$-\frac{(J_i - 1)}{J_i(L+1)}$
$J_f = J_i$	$\frac{-1}{J_i(L+1)}$	$\frac{-1}{2J_i(J_i+1)}$	$\frac{1}{(J_i+1)(L+1)}$
$J_f = J_i + 1$	$\frac{-(J_i+2)}{(J_i+1)(L+1)}$	$\frac{-1}{2(J_i+1)}$	$\frac{1}{(L+1)}$

C. *Mixed transitions*. The asymmetry parameter A is obtained by substituting (5.19) or (5.20) into (5.16″). For the three most important cases given in (5.20),

$$A_l = -\frac{\langle Mh_l \rangle}{(L+1)J_i}[1 - x \pm 2\sqrt{x(1-x)J_i(J_i+1)}\cos\phi] \qquad J' = J_f - L$$

$$A_l = +\frac{\langle Mh_l \rangle}{(L+1)(J_i+1)}[1 - x \pm 2\sqrt{x(1-x)J_i(J_i+1)}\cos\phi] \qquad J' = J_f + L$$

$$A_l = -\frac{\langle Mh_l \rangle}{2J_i(J_i+1)}[1 - x \pm 2\sqrt{x(1-x)J_i(J_i+1)}\cos\phi] \qquad J' = J_f \tag{5.22}$$

(+ sign for e; — sign for ν).

5.5. TRIPLE POLARIZATION CORRELATION (TIME REVERSAL EXPERIMENT)

In mixed transitions, the relative phase of the Fermi and Gamow-Teller components can also be determined by an experiment in which the direction of nuclear polarization and the directions of emission of both the electron and the neutrino are measured. Let us assume that the nuclei are initially polarized in the direction of the positive z-axis, as shown in Fig. 5.6. The electrons are detected at some direction perpendicular to the z-axis, which we can take as the x-axis. The neutrinos are detected in a direction perpendicular to both the direction of nuclear polarization and electron emission; namely in the direction of the y-axis. (It is of course always the direction of nuclear recoil which is observed in practice, not the direction of neutrino emis-

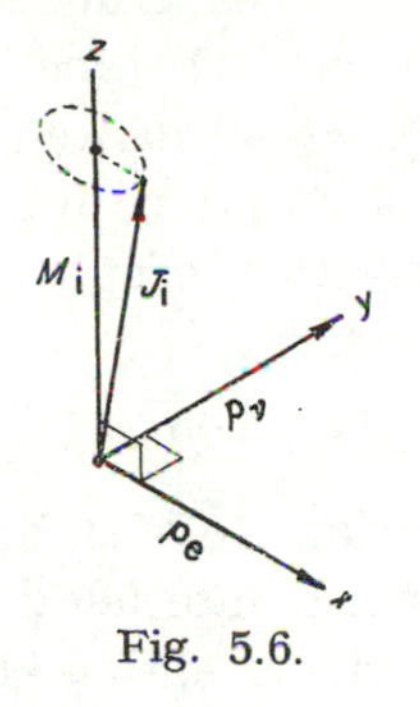

Fig. 5.6.

sion.) The asymmetry is measured, say, between electron emission in the $+x$ and $-x$ directions for a given direction of nuclear polarization and neutrino emission.

Note that reversing the direction of electron emission is the same as reversing the direction of either the nuclear polarization or of the neutrino emission, since there is no preferred direction in space. There are only two possible relative orientations of the three mutually perpendicular directions, M_i, p_e, and p_ν. The three directions either form a left-handed or a right-handed coordinate system, and reversing any one of the directions changes from one to the other. Note also that the character of the coordinate system defined by these vectors does not change under a space inversion ($x \to -x, y \to -y, z \to -z$), because the nuclear polarization is an *axial* vector and remains in the positive z direction after space inversion. Only two of the directions are reversed under space inversion, those of the electron and the neutrino emission. The asymmetry in the direction of electron emission with respect to the directions of nuclear polarization and neutrino emission is invariant under space inversion; therefore parity non-conservation is not required to obtain an effect. This experiment is thus a '*classical*' $1 + A \cos \theta$ experiment like the electron-neutrino angular correlation. (Note that in each of these experiments the directions of emission of both the electron and neutrino are measured.)

If we consider the directions of nuclear polarization and neutrino emission as fixed, this experiment measures the angular distribution of the electrons with respect to the x-axis. Since the electrons are in a state $j_e = \frac{1}{2}$ in allowed transitions, the angular distribution is again $1 + A \cos \theta$ with A given by (4.13). However, θ is now measured with respect to the x-axis and j_x appears in the expression for A instead of j_z,

$$A_x = 2\langle h_e j_{x_e} \rangle . \tag{5.23}$$

To evaluate this expression, we must determine the value of j_x for the particular electron state relevant to this experiment. The wave function describing leptons emitted from a polarized nucleus is given by (4.17)

$$|\text{lept}\rangle = \overset{+}{a}_{GT}|e\uparrow\nu\uparrow\rangle + \overset{-}{a}_{GT}|e\downarrow\nu\downarrow\rangle + \frac{a_{GT} + a_F}{\sqrt{2}}|e\uparrow\nu\downarrow\rangle + \frac{a_{GT} - a_F}{\sqrt{2}}|e\downarrow\nu\uparrow\rangle . \tag{4.17}$$

We cannot use this wave function directly to calculate the asymmetry parameter (5.23). We must first take into account the measurement of the direction of emission of the neutrino. The wave function (4.17) is a linear combination of the two m-states of the neutrino, $\nu\uparrow$ and $\nu\downarrow$. Since in our experiment the neutrino is emitted in the y-direction, it is more convenient to take as our two basic neutrino states the eigenstates of the operator j_{y_ν}, rather than of j_{z_ν}. Even better for our purposes is the operator $h_\nu j_{y_\nu}$, the product of the neutrino helicity and the projection of its angular momentum on the y-axis. This operator also has two eigenvalues, $+\frac{1}{2}$ and $-\frac{1}{2}$. The wave function (4.17) is therefore a linear combination of the two neutrino states having these two eigenvalues of $h_\nu j_{y_\nu}$.

The angular distribution of the neutrinos with respect to the y-axis is also given by a '$1 + A_y \cos \theta$' distribution, with A_y given by (4.13) or (5.23) but referred to the y-axis,

$$P(\theta_y) = 1 + 2\langle h_\nu j_{y_\nu} \rangle \cos \theta_y . \tag{5.24}$$

From (5.24) we note that the probability for neutrino emission into the y-direction ($e_y = 0$) vanishes for $h_\nu j_{y_\nu} = -\frac{1}{2}$. Thus measurement of the neutrinos emitted in the positive y-direction picks out that part of the wave function (4.17) in which neutrinos are emitted in the $h_\nu j_{y_\nu} = +\frac{1}{2}$ state.

The straightforward procedure for calculating the result of the triple polarization correlation experiment is thus as follows: 1) take the lepton wave function (4.17), 2) project out that part of it which has the neutrino in the state $h_\nu j_{y_\nu} = +\frac{1}{2}$, and 3) calculate the electron asymmetry parameter (5.23) with this projected wave function. We can avoid the complications of the projection procedure by the following bit of trickery, using the invariance of the process under rotations.

Let us consider an experiment where the neutrino is detected in the *negative* y-direction. This picks out the $h_\nu j_{y_\nu} = -\frac{1}{2}$ part of the lepton wave function. Such an experiment differs from the preceding case just by a rotation of 180°. The electron angular distribution must then also be rotated by 180°; i.e. the *sign* of the asymmetry must be reversed. Thus

$$A_{x_e}(j_{y_\nu} h_\nu = -\tfrac{1}{2}) = -A_{x_e}(j_{y_\nu} h_\nu = +\tfrac{1}{2}) \,. \tag{5.25}$$

We can calculate the asymmetry parameter for the triple correlation experiment either in the system where $j_{y_\nu} h_\nu = +\frac{1}{2}$ or the one where $j_{y_\nu} h_\nu = -\frac{1}{2}$. Using (5.23) and (5.25), we obtain

$$A = 2\langle h_e j_{x_e} \rangle_{(j_{y_\nu} h_\nu = +\frac{1}{2})} = -2\langle h_e j_{x_e} \rangle_{(j_{y_\nu} h_\nu = -\frac{1}{2})} \,. \tag{5.26}$$

This can also be written

$$A = 4\langle h_e j_{x_e} h_\nu j_{y_\nu} \rangle \,. \tag{5.27}$$

The expectation value in (5.27) can be taken in either of the two neutrino states, since $2h_\nu j_{y_\nu} = \pm 1$ and gives the sign change required (5.26). By putting the quantity $h_\nu j_{y_\nu}$ into the expectation value we have obtained a result which can be used with any neutrino state projected from (4.17) and therefore with the original state (4.17). We thus avoid the necessity of projecting a particular neutrino state out of (4.17).

The expectation value (5.27) can be calculated directly. However, further simplification is possible by noting that a rotation of the experiment by 90° about the z-axis changes j_x into j_y, changes j_y into $-j_x$ and does not change the electron asymmetry, which is then about the y-axis. We thus have

$$A_{x_e}(j_{y_\nu}\, h_\nu = +\tfrac{1}{2}) = A_{y_e}(j_{x_\nu}\, h_\nu = -\tfrac{1}{2}) = -4\langle h_e j_{y_e}\, h_\nu j_{x_\nu}\rangle\,, \quad (5.28)$$

then

$$A = \tfrac{1}{2}\{A_{x_e}(j_{y_\nu} h_\nu = +\tfrac{1}{2}) + A_{y_e}(j_{x_\nu} h_\nu = -\tfrac{1}{2})\} = 2\langle h_e h_\nu (j_{x_e} j_{y_\nu} - j_{y_e} j_{x_\nu})\rangle\,.$$

This can be written

$$A = 2\,\mathrm{Im}\langle h_e h_\nu (j_x + \mathrm{i} J_y)_\nu\, (j_x - \mathrm{i} j_y)_e\,. \quad (5.29)$$

The form (5.29) is particularly suitable for evaluation of the asymmetry parameter in the state (4.17) expressed in terms of the eigenstates (4.15) of j_{z_e} and j_{z_ν}. From (4.14)

$$\langle \nu\!\uparrow e\!\downarrow\!|(j_x + \mathrm{i} j_y)_\nu (j_x - \mathrm{i} j_y)_e|\nu\!\downarrow e\!\uparrow\rangle = 1\,, \quad (5.30)$$

and all the other matrix elements of the operator $(j_x + \mathrm{i} j_y)_\nu$ $(j_x - \mathrm{i} j_y)_e$ vanish. We can now evaluate (5.29) using the wave function (4.17) and the result (5.30). The only non-vanishing contribution is the cross term between the last two terms of (4.17). Thus

$$A = \frac{\mathrm{Im}\langle h_e h_\nu\rangle (a^*_{\mathrm{GT}} - a^*_{\mathrm{F}})(a_{\mathrm{GT}} + a_{\mathrm{F}})}{3|a^2_{\mathrm{GT}}| + |a^2_{\mathrm{F}}|} = \frac{2\langle h_e h_\nu\rangle |a_{\mathrm{GT}} a_{\mathrm{F}}| \sin\phi}{3|a^2_{\mathrm{GT}}| + |a^2_{\mathrm{F}}|}$$

$$= 2 h_e h_\nu \sqrt{x(1-x)}\, \sin\phi\, \sqrt{J_i/(J_i+1)}\,. \quad (5.31)$$

The subscripts and arguments of A have been omitted.

From (5.13) we note that the asymmetry is proportional to the mean *relative* helicity $\langle h_e h_\nu\rangle$ and is indeed not dependent upon parity non-conservation effects. It is proportional to $\sin\phi$ rather than to $\cos\phi$ and is therefore more sensitive to small deviations from 180° than the lepton angular distribution interference experiments, which give effects proportional to $\cos\phi$. On the other hand, if both amplitudes are real, there is no asymmetry.

The maximum possible asymmetry is not unity, but $\sqrt{J_i/(J_i+1)}$, because all the asymmetry comes from the interference term. Unlike the lepton angular distribution experiment, there is no asymmetry in a pure Gamow-Teller transition and the $m_t = \pm 1$ components do not contribute.

For the case of the neutron decay, where other experiments indicate that $\phi = 180°$, this experiment is the most sensitive way to determine whether there are small deviations from 180°. The experimental results show no asymmetry within experimental error and indicate that ϕ is indeed 180°, and therefore that the two amplitudes a_F and a_{GT} can be taken as real. This is interpreted according to beta decay theory as indicating invariance under time reversal.

The relation between this triple correlation experiment and time reversal invariance can be qualitatively understood by noting that all three directions measured in the experiment change sign under time reversal, the direction of nuclear polarization, the direction of the electron momentum, and the direction of the neutrino momentum. Thus the direction of the asymmetry would be reversed under a time reversal transformation. This kind of argument is not rigorous, however, because time reversal also changes outgoing waves to incoming waves and this effect invalidates the general argument when effects of the Coulomb field of the residual nucleus must be taken into account. In the neutron decay these Coulomb effects are negligible, because of the low value of the proton charge.

Because of the implications of the neutron triple correlation experiment regarding time reversal invariance, this experiment is sometimes called the 'time reversal experiment'.

CHAPTER 6

CONNECTION WITH BETA DECAY THEORY. CONCLUSIONS

6.1. GENERAL

In the preceding chapters we have seen the extent to which allowed transitions in beta decay can be described without use of beta decay theory. We have shown how complete sets of experiments can be chosen which give all possible independent information about the leptons emitted in a particular decay; i.e. which determine completely the amplitudes for the different channels (2.8). We have also noted that specification of these amplitudes is the most convenient way to express the predictions of the theory. It is now of interest to examine the predictions of beta decay theory from this point of view. We shall not go into the details of the theory, as this is beyond the scope of the present treatment. Instead, we shall merely quote some of the relevant results of the theory without derivation.

The presently accepted theory of beta decay does not give unique predictions of experimental results. There appear in the theory a number of arbitrary parameters, the so-called *coupling constants* whose values are not predicted by the theory. The values of the lepton amplitudes predicted by the theory are functions of these coupling constants. There are therefore two kinds of relations between lepton amplitudes which are predicted by beta decay theory:

1. Relations independent of the values of the coupling constants. Experimental verification of these relations tests the validity of the theory of beta decay.

2. Relations depending upon the values of the coupling constants. If the validity of beta decay theory is assumed, then experimental investigation of these relations serves to determine the values of the coupling constants.

In the remainder of this chapter we express the predictions of beta decay theory in terms of the lepton amplitudes (2.8) and show that there is a natural division into the two categories defined above. We can then re-examine our complete set of experiments and define a smaller set which determines the lepton amplitudes completely *if we include an additional assumption that beta decay theory is valid:* i.e. that the relations of type (1) above hold between the lepton amplitudes. If we take into account also the practical limitations on the accuracy of the experiments in beta decay, we can then give a realistic evaluation of the experimental determination of the lepton amplitudes, and therefore also of the coupling constants appearing in beta decay theory.

Because the effect known as Fierz interference is important in predicting useful relations between the lepton amplitudes, a detailed discussion of the effect is given in section 4. Although this properly belongs in the domain of beta decay theory, the essential physical features can be presented in terms of the lepton channels and amplitudes without going into the details of the theory. In addition, the point of possible additional degrees of freedom for the neutrino, which was left open in Chapter 2, § 2, is discussed.

6.2. RELATION OF BETA DECAY THEORY TO THE SPECIFICATION OF LEPTON STATES

We have seen that a complete description of the experimental results in allowed beta decay is given in terms of eight complex quantities specifying the probability amplitudes for the decay into the eight independent lepton channels. The numerical values of these amplitudes are to be determined by experiment. These results have been obtained without any explicit reference to the theory of beta decay. They are therefore independent of the particular details of beta decay theory.

It is now of interest to see what additional information and insight into the beta decay process is obtained from a more detailed fundamental theory. We should like to know, for example, whether the theory gives explicit relations for the values of the eight independent lepton amplitudes. We shall not develop this theory here as it is beyond the scope of this book and is treated extensively elsewhere. We merely quote the results regarding the *kind of information* available from such a theory.

Fermi's original formulation of beta decay allows for five different covariant interactions. The strengths of these interactions appear as free parameters in the theory, the coupling constants. The five interactions are called scalar, vector, tensor, axial vector and pseudoscalar (S, V, T, A, P), because of their transformation properties under Lorentz transformations. The five complex coupling constants are usually denoted by C_S, C_V, C_T, C_A, and C_P. In allowed transitions, there is no contribution from the pseudoscalar interaction; there remain only four relevant coupling constants.

The original theory assumed Lorentz and space inversion invariance and a local interaction. Relaxation of the requirement of space inversion invariance to allow for parity non-conservation doubles the number of possible interactions and therefore of coupling constants. The five additional interactions were first described by the use of primed coupling constants (C'_S, C'_V, C'_T, C'_A and C'_P). It is also possible to consider linear combinations of the corresponding primed and unprimed interactions to obtain interactions in which left-handed or right-handed neutrinos are emitted exclusively. In this description the eight coupling constants relevant to allowed beta decay are written with superscripts indicating the neutrino helicity:

$$C_S{}^L, C_V{}^L, C_T{}^L, C_A{}^L, C_S{}^R, C_V{}^R, C_T{}^R, C_A{}^R \, .$$

This description is more convenient for our purposes.

Beta decay theory describes allowed beta decay in terms of eight complex quantities specifying the strengths of the different possible interactions. The numerical values of these constants must be determined from experiment. Thus it appears at first that the detailed beta decay theory tells us no more about allowed beta decay than we knew in the first place from general considerations about the number of possible decay channels. The description of the experiments requires the determination of eight complex constants which must be determined by experiment! Beta decay theory does not give us any further information about these constants for any particular case!

However, it is only when we limit ourselves to a particular decay and to a particular electron energy that we can learn nothing more from beta decay theory than we already know from our lepton amplitudes. The lepton amplitudes are unknown functions of the electron energy and may vary in an arbitrary manner from one decay to another. The beta decay coupling constants, on the other hand are *universal constants* applicable to all beta decay processes and electron energies. Thus, once we have determined these eight coupling constants we can in principle calculate the probability amplitudes for all lepton channels in allowed beta decay for all cases where the nuclear wave functions are known. Forbidden transitions can also be calculated, if the two additional pseudoscalar coupling constants are known.

Although beta decay theory cannot predict a priori the relative magnitudes of the lepton amplitudes *for a given transition and electron energy*, the theory gives the following additional information about the lepton amplitudes:

1. *The dependence of the lepton amplitudes upon the energy of the emitted electron.* This gives the shape of the beta decay electron spectrum and the energy dependence of the electron polarization.

2. *The variation of the lepton amplitudes from one decay to another as a function of the structure of the nuclear states involved.* Predictions, however, are not easily made because of uncertainties in the structure of *nuclei* and in the structure of the *nucleon*. The nuclear structure effects can be estimated in cases where a particular nuclear model is known to apply, such as the shell model or the unified model, and where the configuration is well

established. In general the structure of the nucleus is not understood as well as the beta interaction, and the comparison of experiments with theory gives information about the nuclear structure rather than the beta interaction.

A fundamental difficulty in any relativistic beta decay theory is the absence of a satisfactory relativistic decription of the nucleon and its strong interactions. The original Fermi theory treats the nucleon as an ordinary Dirac particle. It has been suggested, for example, that the strong nucleon-pion interaction may give rise to effects in beta decay analogous to those in electrodynamics believed responsible for the anomalous magnetic moments of the nucleons (weak magnetism; GELL-MANN [1958]). These effects are always small, of the same order of magnitude as the contribution of the next higher order of forbiddenness to a given transition, and are difficult to detect experimentally.

Although uncertainties in nuclear structure make difficult the prediction of the variation of lepton amplitudes from one decay to another, there is one case where these uncertainties are irrelevant and unique predictions can be made. This is the case where a given coupling constant is zero; i.e. the particular interaction is absent. In such a case, all contributions to the lepton amplitudes which would be proportional to this coupling constant vanish for all decays irrespective of the structure of the nuclei. Fortunately, this case seems to obtain for nearly all the coupling constants; only two seem to be non-zero in the light of present experimental information. Thus it is possible to obtain precise information about beta decay from decays of complex nuclei without having detailed knowledge of the structure of the nuclei (and vice versa).

If a universal Fermi interaction is assumed, as seems reasonable in the light of the experimental data, the processes of the μ-meson decay and μ-meson capture in nuclei can also be treated.

3. *The classification of transitions as allowed or forbidden in various degrees.* The qualitative argument given in Chapter 2,

§ 4, restricting lepton angular momenta to the smallest possible values is put on a firm rigorous basis, and the small contributions of the higher angular momentum channels can be calculated quantitatively. Selection rules involving the *parities* of the nuclear states are also obtained and are valid *even though parity is not conserved in the decay*. The strong interactions responsible for nuclear forces are parity conserving, and the nuclei are to a very good approximation in states of definite parity (the mixing of parities due to the beta decay interaction is negligible because of the weakness of the beta decay interaction). A parity selection rule is obtained for allowed transitions; namely that *there is no change in parity between the initial and final nuclear states.*

4. *Direct relations between the three types of beta decay processes, negaton emission, positon emission and electron capture.* These follow from the description of the leptons by the Dirac equation and from the locality of the interaction. The relations are between processes in which the *charge and the helicity* of all leptons are reversed. Note that this has nothing to do with the transformation of *charge conjugation,* which would require the transformation of *nuclei* to *anti-nuclei* as well. The relations are valid independently of the validity of charge conjugation invariance. In fact they are valid even though charge conjugation is known to be *violated* in beta decay. The result can be stated as follows:

The predictions of beta decay theory for the amplitude of a particular lepton channel in positon emission are related to those for the corresponding channel in electron capture with the helicity of the electron reversed and the helicity of the neutrino unchanged, and the corresponding channel in negaton emission with the helicities of *both leptons* reversed. The exact quantitative specification of this relation is not possible without direct reference to the theory because differences in the structure of the relevant nuclei are significant and are not treated in our simple formulation. However, if the amplitude for a particular channel *vanishes* for one type of decay, it then vanishes for the corresponding channel in all three types of decay.

In the R-L notation for the beta decay coupling constants described above (C_S^R, C_S^L, etc.) it is customary to use the letters R and L as referring to the helicity of the neutrino emitted in *positon decay*. A coupling constant with the superscript R therefore describes an interaction in which *right*-handed neutrinos are emitted in *positon* decay and *electron capture*, and *left*-handed neutrinos are emitted in *negaton* decay.

With these definitions, experimental evidence at present indicates that all of the coupling constants vanish except two, C_V^L and C_A^L. The neutrino emitted with positons and in electron capture is always left-handed, and the neutrino emitted with negatons is always right-handed.

6.3. THE RELATION BETWEEN BETA THEORY INTERACTIONS AND LEPTON AMPLITUDES IN ALLOWED DECAY

Let us now examine how the eight complex amplitudes (2.8) specifying allowed beta decay into the various channels are determined from beta decay theory as functions of the eight beta decay coupling constants and the electron energy. The relation for extremely relativistic electron velocities ($v_e \approx c$) turns out to be particularly simple. There is a one to one correspondence between the eight beta decay interactions and coupling constants and the eight amplitudes for the channels illustrated in Fig. 3.2. The S and V interactions contribute only to Fermi transitions, the T and A only to Gamow-Teller. The two leptons are emitted with the *same* helicity ($h_{rel} = +1$) in the S and T interactions, with *opposite* helicity ($h_{rel} = -1$) in the V and A interactions. The results are summarized in Fig. 6.1.

The experimental results seem to indicate that only C_V^L and C_A^L are not zero. The leptons are emitted with opposite helicity in both Fermi and Gamow-Teller transitions. The positon is right-handed, emitted with a left-handed neutrino, the negaton is left-handed, emitted with a right-handed neutrino. Full interference is observed between the Fermi and Gamow-Teller amplitudes, since all helicities are the same in both transitions.

At lower electron energies both theory and experiment show that the electron helicity and the relative helicity are no longer 100 % for any given interaction. The Fermi-Gamow-Teller classification, however, remains good for all electron energies, as is

GT	F	
T	S	$h_e = h_\nu$
A	V	$h_e = -h_\nu$

Fig. 6.1.

the helicity of the neutrino. Thus, in general, each beta interaction contributes to two decay channels if they are defined according to (2.8), namely the one which takes the entire contribution in the extreme relativistic limit and the one in which the electron helicity is reversed, but all other quantum numbers remain the same. The second channel can also be called the one in which the *relative* helicity is reversed, since reversing the relative helicity with a fixed neutrino helicity is the same as reversing the electron helicity. The mixing of the two channels for each individual beta interaction turns out to be such that the magnitude of the average electron helicity $\langle h_e \rangle$ is just $|\langle h_e \rangle| = v_e/c$.

Fig. 6.2 shows the contributions of the different interactions

GT	F	
$\frac{1+v/c}{2}$ T; $\frac{1-v/c}{2}$ A	$\frac{1+v/c}{2}$ S ; $\frac{1-v/c}{2}$ V	$h_e = h_\nu$
$\frac{1+v/c}{2}$ A; $\frac{1-v/c}{2}$ T	$\frac{1+v/c}{2}$ V ; $\frac{1-v/c}{2}$ S	$h_e = -h_\nu$

Fig. 6.2.

to the lepton channels at any electron velocity. The relative magnitude of the contribution of each interaction to the various channels is given by the factor $\frac{1}{2}(1 \pm v/c)$ written in front of this interaction. Thus, for example, the contribution of the S interaction is divided between the two Fermi channels of opposite relative helicity with probabilities $\frac{1}{2}(1+v/c)$ for $h_e = h_\nu$, and $\frac{1}{2}(1-v/c)$ for $h_e = -h_\nu$.

The variation of the electron helicity with energy can be understood qualitatively as follows. The helicity of an electron is not Lorentz-invariant; its value depends upon the velocity of the observer relative to the electron. In the rest frame of the electron the helicity is completely undefined as the momentum is zero. In two frames moving in opposite directions with respect to the rest frame, the electron velocity will be in opposite directions and the helicities in these two frames will be measured to be opposite to one another. The helicity is defined in a relativistically invariant way only for a zero mass particle like the neutrino, which is always moving with velocity c and which has no rest frame. Thus the helicity of the neutrino plays a more fundamental role in beta decay than the electron helicity. A simple description of the electron helicity occurs only in the extreme relativistic region, where the rest mass of the electron can be neglected and it behaves approximately like a zero mass particle.

At zero electron energy, there is no direction defined for the momentum and therefore no preferred direction for the spin. One might expect the average electron helicity to go to zero at low energies. We have already seen in Chapter 2, § 6, equation (2.10) that this is indeed the case in the non-relativistic limit.

One might ask whether it is possible to change the definition of our beta decay lepton channels in such a way that there would be a one to one correspondence between the channels and the beta decay interactions at all energies, instead of only in the extreme relativistic limit. We might try to find a relativistic generalization of the relative helicity operator h_{rel} to specify the channels. To find such an operator, we should examine the properties of the wave functions describing the emission of electrons in *pure* S, T, V, and A interactions. An examination of these wave functions in the non-relativistic limit is sufficient to show that no suitable operator exists.

In the non-relativistic limit, $v_e = 0$, the orbital angular momentum becomes a constant of the motion, and electrons in allowed decays are emitted in the $s_{\frac{1}{2}}$ state, as was pointed out in Chapter 2, § 4. There is only this single electron state relevant to allowed beta decay, rather than two states. Thus in allowed Fermi transitions, both the S and V interactions must lead to the same state, and in allowed Gamow-Teller transitions both the V and A interactions must lead to the same state. If we try to define our Fermi channels by the properties of wave functions of electron emitted in the S and V interactions, these wave functions have opposite helicities in the extreme relativistic limit, but are identical in the *non-relativistic* limit. They cannot therefore be always *orthogonal* to one another, particularly in the non-relativistic limit. They therefore cannot be eigenfunctions of some generalized relativistic helicity operator with different eigenvalues. The same argument applies to the Gamow-Teller channels.

If we defined our lepton channels according to the SVTA interactions instead of by the relative helicity, our basic states would not be orthogonal but could be linearly independent except in the non-relativistic limit. Because of the non-orthogonality, the *total decay probability* would not be given by the sum of the squares of the channel amplitudes; there would also be cross terms due to the overlap of the basic wave functions. The total decay probability in Fermi transitions expressed in terms of the S and V interactions thus would contain *energy dependent* cross terms, which would vanish in the extreme relativistic limit where the S and V wave functions are orthogonal. Such cross terms would therefore be detectable in measurements of the variation of the total decay probability with energy; i.e. by measurements of the shape of the continuous beta ray spectrum, as well as in other measurements. These cross terms are known as Fierz interference terms: their effects are considered in detail in the following section.

We see therefore that there is no simple way to define lepton channels by the beta decay interactions, and we may as well stick with our relative helicity definition which gives energy-dependent amplitudes. We need only remember that beta decay theory prescribes exactly the energy dependence of these amplitudes, so that a measurement of the amplitudes at a single electron energy is sufficient to determine all the free parameters relevant to the particular decay occurring in the theory.

6.4. FIERZ INTERFERENCE

Let us now examine the phenomenon of Fierz interference in more detail. The simplest type of Fierz interference occurs between two channels having opposite electron helicity and the same values for all other quantum numbers. The wave functions corresponding to these two orthogonal states can be represented as orthogonal vectors in a two dimensional Hilbert space as shown in Fig. 6.3. Let us choose the vertical and horizontal axes

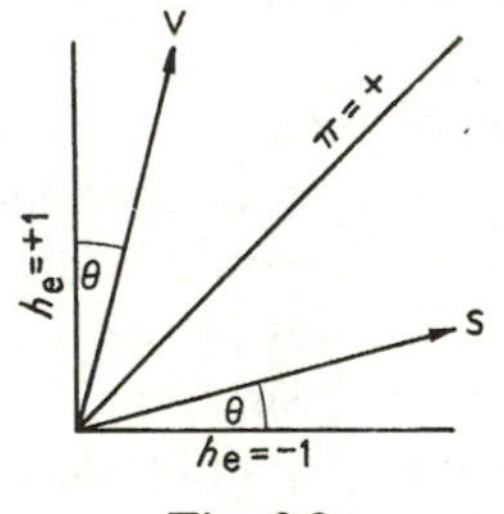

Fig. 6.3.

as the directions of vectors representing the states of positive and negative electron helicity respectively. A vector at an angle θ with respect to the vertical ($h_e = +1$) axis is a mixture of states having positive and negative helicities with amplitudes proportional to $\cos\theta$ and $\sin\theta$. The mean helicity in such a state is then

$$\langle h_e \rangle_\theta = \cos^2\theta - \sin^2\theta = \cos 2\theta \,. \tag{6.1}$$

A vector at an angle $\theta = 45°$ represents a state which is an equal mixture of both helicities and has mean helicity zero. In the non-relativistic limit electrons are emitted in an even-parity s-state which is a linear combination of the two helicity states with equal amplitude and phase. This state is represented in Fig. 6.3 by the vector at an angle of 45° labeled $\pi = +$.

Let us consider the case of a pure Fermi transition in which a left-handed neutrino is emitted. (Experimental results indicate that this is the case in *positon* decay.) Then in the extreme relativistic limit, the V interaction gives right-handed electrons ($h_e = +1$) and the S interaction gives left-handed electrons ($h_e = -1$). In the non-relativistic limit both interactions lead to the same state; namely that represented by the vector $\pi = +$ at 45°. Thus as the electron velocity decreases, the vectors representing the electron states for the V and S interactions rotate toward one another, as shown in Fig. 6.3. At some intermediate velocity v, the V interaction gives a mean helicity $+v/c$; the S interaction gives a mean helicity $-v/c$. From (6.1) we see that the electron state for the V interaction is represented by a vector at an angle θ with respect to the *vertical axis*, given by

$$\cos 2\theta = \frac{v}{c}. \tag{6.2}$$

The electron state corresponding to the S interaction is represented by a vector making the same angle θ (6.2) with the *horizontal* axis. Thus the V and S vectors for a given electron velocity are always symmetric about the line $\pi = +$ at 45°. Note that the V and S vectors are not orthogonal to one another.

Let us now consider a particular case at some velocity v in which both the S and V interactions are present, with amplitudes a_S and a_V, as shown in Fig. 6.4. The amplitudes are assumed to be real, in order to allow them to be represented simply on a vector diagram. However, the general case of complex amplitudes does not differ in principle from the real case. Rather the results with an arbitrary relative phase between the two amplitudes are intermediate between the two extreme cases of 0° and 180° shown in Fig. 6.4.

The state of the electron emitted in this mixed S and V transition is represented by a vector $\boldsymbol{a}$ in Fig. 6.4 which is the vector sum of the vectors $\boldsymbol{a}_S$ and $\boldsymbol{a}_V$. The *magnitude* of the vector $\boldsymbol{a}$ is related to the overall decay probability; the *direction* of the vector $\boldsymbol{a}$ is related to the mean electron helicity by the relation (6.1). The principal results of Fig. 6.4 can be summarized qualitatively as follows:

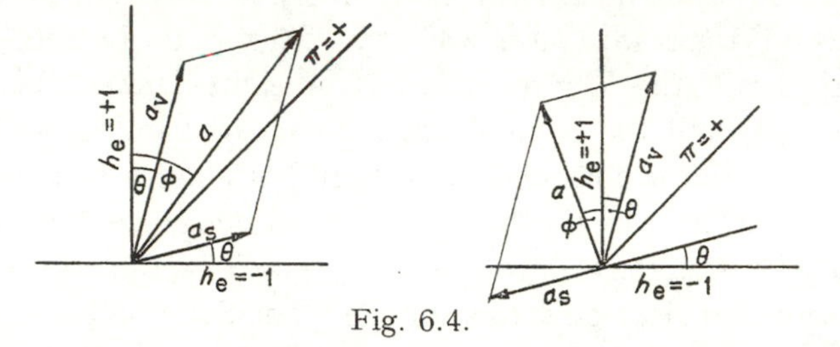

Fig. 6.4.

1. Because the two vectors $\boldsymbol{a}_S$ and $\boldsymbol{a}_V$ are not orthogonal, the square of the vector $\boldsymbol{a}$ may be either larger or smaller than the sum of the squares of the vectors $\boldsymbol{a}_S$ and $\boldsymbol{a}_V$, depending upon the relative phase. The total decay probability can therefore be greater or less than the sum of the individual contributions of the S and V interactions. The difference is zero in the extreme relativistic limit, where $\boldsymbol{a}_S$ and $\boldsymbol{a}_V$ are orthogonal, and increases with decreasing electron velocity. The result of this effect would be observed as an additional energy dependence in the decay probability in addition to that expected for a pure S or pure V interaction; i.e. the electron energy spectrum would have a different shape.

2. The mean electron helicity can be either less than or greater than v/c, depending upon the relative phase of the vectors $\boldsymbol{a}_S$ and $\boldsymbol{a}_V$. There is no reason to be surprised at values of the mean helicity which are greater than v/c, as there is no particular physical principle restricting the polarization to be less than v/c.

These two results can be expressed quantitatively by calculating the magnitude and the orientation of the vector $\boldsymbol{a}$ from the vector diagram, Fig. 6.4. The square of the magnitude of $\boldsymbol{a}$ is given by

$$\begin{aligned} a^2 &= a_V{}^2 + a_S{}^2 + 2a_S a_V \sin 2\theta \\ &= (a_V{}^2 + a_S{}^2)\left(1 + \frac{2a_S a_V \sin 2\theta}{a_S{}^2 + a_V{}^2}\right). \end{aligned} \tag{6.3}$$

Let ϕ be the angle between the vector $\boldsymbol{a}$ and the vertical ($h_e = +1$) axis. Then we have

$$\sin\phi = \frac{1}{a}(a_V \sin\theta + a_S \cos\theta)$$
$$\cos\phi = \frac{1}{a}(a_V \cos\theta + a_S \sin\theta)\,. \tag{6.4}$$

Then

$$\cos 2\phi = \cos^2\phi - \sin^2\phi = \frac{1}{a^2}(a_V{}^2 - a_S{}^2)\cos 2\theta$$
$$= \cos 2\theta \, \frac{(a_V{}^2 - a_S{}^2)}{(a_V{}^2 + a_S{}^2)} \Big/ \left(1 + \frac{2a_S a_V \sin 2\theta}{a_S{}^2 + a_V{}^2}\right). \tag{6.5}$$

Let us now use the results (6.3) and (6.5) to obtain expressions relating the physical quantities v_e, h_e, and the probability $W(v_e)$ for emission of an electron of velocity v_e. From (6.2) we obtain

$$\sin 2\theta = \sqrt{1 - (v_e{}^2/c^2)} = \frac{m}{E} \tag{6.6}$$

where E is the total energy (kinetic plus rest energy) of the electron.

The probability $W(v_e)$ is proportional to a^2, (6.3). This can be expressed in terms of the electron energy E using (6.6). Thus

$$W(v_e) \propto (a_V{}^2 + a_S{}^2)\left(1 + 2\gamma b\, \frac{m}{E}\right). \tag{6.7}$$

The mean helicity is obtained by substituting (6.6) into (6.1),

$$\langle h_e \rangle = \frac{v_e}{c} \frac{(a_V{}^2 - a_S{}^2)}{(a_V{}^2 + a_S{}^2)} \cdot \frac{1}{1 + 2\gamma b m/E}, \tag{6.8}$$

where

$$\gamma b = \frac{a_S a_V}{a_S{}^2 + a_V{}^2}. \tag{6.9}$$

The characteristic Fierz interference factor, $1 + 2\gamma b(m/E)$, appears both in the expression for the decay probability $W(v_e)$ and in the expression for the mean helicity $\langle h_e \rangle$. As is to be expected, the effect increases with decreasing energy, as the S and V wave functions become less orthogonal and overlap more.

For the general case where the amplitudes are complex, the same analysis can be carried through, and the results are given by (3.24) and (3.25) with γb defined as

$$\gamma b = \frac{\mathrm{Re}(a_S{}^* a_V)}{(|a_S{}^2| + |a_V{}^2|)}.$$

$$= \frac{|a_S a_V| \cos \eta}{(|a_S{}^2| + |a_V{}^2|)}, \tag{6.10}$$

where η is the phase angle between the S and V amplitudes.

The parameter (6.10) is called γb here in order to conform with the standard notation in beta decay theory. The separation into two factors γ and b becomes convenient when (6.10) is expressed in terms of the *coupling constants* C_S and C_V of the two interactions, instead of the amplitudes a_S and a_V. Such a calculation is beyond the scope of our treatment; the results indicate that the amplitudes are proportional to the appropriate coupling constants, as might be expected, but the proportionality factor is complex. There is a phase factor introduced into the amplitudes as a result of the interaction of the electron with the Coulomb field of the nucleus. Thus the *amplitudes* a_S and a_V are in general complex, even though the coupling constants are probably real (as would be required by time reversal invariance of the theory). It is therefore convenient to split the phase shift η into two parts, the phase shift η_{SV} between the two coupling constants, and the Coulomb phase shift η_C. We thus have

$$\cos \eta = \cos \eta_C \cos \eta_{SV} - \sin \eta_C \sin \eta_{SV}. \tag{6.11}$$

When the coupling constants are real, as required by time reversal invariance, $\sin \eta_{SV}$ is zero and the second term in (6.11) vanishes. The factor γ in (6.10) is then just $\cos \eta_C$.

Similar Fierz interference terms appear in the Gamow-Teller transitions. The analysis is exactly the same as in the Fermi treatment except that a_T and a_A replace a_S and a_V. In mixed transitions, more complicated Fierz terms can appear as combined Fierz-Gamow-Teller interference between the S and A interactions and between V and T. This can be seen from the result (5.14) for the interference effect between the Fermi and

Gamow-Teller amplitudes in mixed transitions. The effect is proportional to $a_{GT}a_F \cos \phi$. If a_F is a linear combination of a_S and a_V, as shown in Fig. 6.4 and a_{GT} is a similar linear combination of a_T and a_A, the product $a_{GT}a_F$ is the sum of four different interference terms.

Fierz interference has never been observed experimentally. Since the absence of Fierz interference implies the absence of certain products of amplitudes, important conclusions can be drawn regarding the absence of certain interactions. However, we note that if electrons are fully polarized in the extreme relativistic limit, as seems to be the case, then there can be no Fierz interference. Full polarization in the extreme relativistic limit means that *for a given neutrino helicity* only one of the two vectors in Fig. 3.6 is present (in the Fermi case, that only one of the interactions S and V is present). There can therefore be no Fierz interference at lower energies.

The experimental data at present indicate that only the V and A interactions are present. The only possible interference is therefore that between V and A, which is the Fermi-Gamow-Teller interference described in Chapter 3, § 3. However, we shall see in § 7 of this chapter that the absence of Fierz interference is still significant, even though it leads to no conclusions additional to those which already follow from 100 % electron polarization. This is because the precision of the Fierz interference experiments is considerably higher than that of the electron polarization measurements.

6.5. LEPTON CONSERVATION AND THE TWO AND FOUR COMPONENT THEORIES

Let us now return to the question posed in Chapter 2, § 2 regarding the completeness of our set of eight quantum numbers and whether the two helicity states describe all the internal degrees of freedom of the neutrino. If the neutrino is described by a Dirac equation like the electron, but in which the mass and charge are both set equal to zero, the neutrino is described by a four component spinor having four states for a given set of total momentum and angular momentum. For each of the two helicity values, there are two states corresponding to the negaton and the positon, but having essentially the same properties because the neutrino has no electric charge. One might interpret them as particle and anti-particle by analogy with electrons, and postulate a 'lepton conservation' law requiring that leptons be created only in particle-anti-particle pairs.

There are, however, possibilities for describing a neutrino in a relativistic way with only two states, because of its zero mass and charge, provided that a conservation law is sacrificed. One possibility, stemming from the zero charge and sacrificing lepton conservation was proposed by MAJORANA [1937]. Another possibility, stemming from the zero mass and sacrificing parity conservation is the presently accepted two component theory of LEE and YANG [1957], SALAM [1957] and LANDAU [1957].

We shall not go into the details of these theories here, but will simply ask how these considerations effect our treatment in terms of specification of the lepton final states. The question of the adequacy of our treatment hinges upon the answers to the following two key questions:

1. Are more than two internal states required to describe the neutrino?

2. Does there exist a conservation law restricting the emission of neutrinos; i.e. are the neutrinos emitted in negaton decay different from those emitted in positon decay?

Since there are two questions, each of which can be answered independently yes or no, we have four possible theories, known commonly as four or two component theories, (according to the answer to the first question) with or without lepton conservation (according to the answer to the second).

Let us first see how an affirmative answer to the first question would affect our previous analysis. Let us suppose that there are two kinds of neutrinos for each helicity state. Call these for convenience, ν_{L1}, ν_{L2}, ν_{R1} and ν_{R2}. The number of decay channels would then be doubled, since for each case in our previous treatment there would be two, one with ν_1 and one with ν_2. The total number of physically significant independent amplitudes would, however, be less than double the previous number, because of ambiguities in defining ν_1 and ν_2. The two neutrino states of the same helicity are indistinguishable from one another, except for their beta decay interaction. There is no physical reason for calling any one particular state ν_1 or ν_2. Any two linear combinations of the two states which are orthogonal to one another can be picked as the basic neutrino states ν_1 and ν_2 and we can choose them to suit our convenience. A linear transformation in the space of the two states of a given helicity which changes the state which we call 1 and 2 will not change any physical result. (It will of course change the amplitudes for the decay channels labeled 1 and 2 and the coupling constants of the beta decay theory labeled 1 and 2 in accordance with this transformation.) The formal properties of these transformations in beta decay theory were developed by PAULI [1957]. We shall not consider them further, but simply note that they enable us to reduce the number of independent amplitudes or coupling constants by choosing the '1' neutrino to be the one emitted in the V interaction. We thus set the amplitude of the '2' neutrino to be zero for all channels associated with the V interaction.

Since the two neutrino states of the same helicity would be distinguishable only by their beta interaction, the existence of two independent different neutrino states of the same helicity could only be detected in the following two ways:

1. By inverse beta decay experiments. Neutrinos emitted in a given channel in a given decay would be used to excite inverse beta decay in a different channel of the same decay or in a different decay. If different neutrino states of the same helicity were emitted in different types of decays, certain of these combined decay-inverse-decays processes would be forbidden or hindered because the neutrino emitted in the decay would be mainly the wrong type to initiate the inverse decay.

2. By double beta decay experiments. Double beta decay can be observed if a transition from one nuclear state to another

differing from it by two units of charge is energetically possible with the emission of two negatons or the capture of two orbital electrons, while the single beta decay to the intermediate state is energetically impossible. As this decay can be considered as a succession of two virtual single beta decays, the question arises as to whether the neutrino emitted in the first decay is of the right kind to be absorbed in the second. If it is double beta decay is possible *without the emission of neutrinos*; otherwise two neutrinos must be emitted in double beta decay.

3. By interference experiments. If two neutrinos of the same helicity but of different types are emitted into different channels in the same decay, no interference is observable between the two channels. No feasible measurement exists which can mix coherently the two types of neutrinos. For example, if the neutrinos emitted in allowed Fermi transitions were different from those emitted in allowed Gamow-Teller transitions, interference would be reduced. If the two neutrino states were orthogonal, the interference would be zero.

The two key questions can be answered very easily because experiments show that so many of the beta interaction coupling constants vanish. Since only C_V^L and C_A^L are non-zero, the only possibility for more than one right-handed neutrino state to occur in beta decay could be if the right-handed neutrino emitted in allowed Gamow-Teller negaton decays were different from that emitted in allowed Fermi negaton decays. However, experiments show that the full interference effect is observed in mixed transitions. This indicates that there can be only one right-handed neutrino emitted with negatons and only one right-handed neutrino emitted with positons. There are therefore only two states of the neutrino encountered in beta decay and the two component theory is valid.

The question of lepton conservation thus becomes trivial in beta decay. Since the neutrinos emitted in negaton decay all have the same helicity and have helicity opposite to those emitted in positon decay, there is a definite difference between neutrinos emitted in positon and negaton decay, without invoking an additional conservation law. It is irrelevant whether one is called a particle and the other an antiparticle and lepton conservation is required. Any transitions forbidden by lepton conservation (such as double beta decay without neutrino emission) would already be forbidden by helicity selection rules. The neutrino emitted in the first virtual stage of double beta decay has the wrong *helicity* to be *absorbed* in the second. No new theoretical predictions are obtained by postulating that it also has the wrong 'lepton charge'.

Lepton conservation becomes significant when processes involving μ-mesons are also considered. The beta decay results indicate that one can define a lepton charge consistent with all experimental results. The negaton and the left-handed neutrino have the same lepton charge. Whether one calls this pair the *particles* or the *anti-particles* is arbitrary; they are commonly chosen to be *particles*. The other pair, the positon and the right-handed neutrino, have the opposite lepton charge and are commonly taken as *anti-particles*.

In the decay of the μ-meson into an electron and two neutrinos, the results of the electron energy spectrum indicate that the two neutrinos have opposite helicity; i.e. opposite lepton charge. If it is assumed that the neutrinos emitted in muon decay are the same as those emitted in beta decay, this result and lepton conservation fixes the assignment of the muon lepton charge. The positive muon has the same lepton charge as the positon, the negative muon as the negaton, since the neutrino pair emitted in the decay carry no net lepton charge.

If the neutrino emitted in the decay of the pion into a muon and a neutrino is assumed to be the same as the beta decay neutrino, its helicity is determined by the requirement of lepton conservation. The pion is not a lepton and has no lepton charge, the two leptons emitted in pion decay must have opposite lepton charge. The positive pion must decay into a positive muon and a left-handed neutrino; a negative pion into a negative muon and a right-handed neutrino. The same is true for K-meson decay into a muon and neutrino. Similar considerations predict the helicity of the neutrino emitted in the capture of muons by nuclei. The neutrino must have the same lepton charge as the absorbed muon. A right-handed neutrino must be emitted in the capture of a positive muon, a left-handed neutrino must be emitted in the capture of a negative muon. These predictions constitute non-trivial checks on the validity of lepton conservation: they predict neutrino helicities which are not already fixed by other considerations. Since experimental results thus far are in agreement with these predictions, it seems reasonable to assume that the concept of lepton conservation is a useful one and is valid in processes involving electrons, muons and neutrinos.

In all these discussions we have consistently used the terms 'right-handed' and 'left-handed' for the two neutrino states, rather than the terms neutrino and anti-neutrino. The former denote measurable physical properties of the neutrinos, whereas the latter designation involves an arbitrary choice of which particles are called *particles* and which are called *anti-particles*.

Lepton conservation would have a non-trivial meaning in beta decay only if neutrinos of the *same helicity* were emitted in both positon and negaton decay. Lepton conservation would then require that these two neutrinos have opposite lepton charge; i.e., that they be different particles. There would therefore have to be two neutrinos for each helicity state and four states for the neutrino.

We now can understand clearly the relations between the possibilities of a two component theory of the neutrino, the laws of conservation of parity and lepton charge and of invariance under charge conjugation, and the zero mass and charge of the neutrino. If neutrinos of the *same helicity* are emitted in *both positon and negaton* decays, we must give up lepton conservation to have a two component theory as the same neutrino must be emitted in both positon and negaton decays. This is possible without violating charge conservation only if the neutrino has zero charge. On the other hand, to have a two component theory *with lepton conservation*, we must require that the neutrinos emitted in positon and negaton decays be different, i.e., have opposite helicities. This clearly violates both charge conjugation and parity conservation, as it correlates a definite helicity with a sign of charge. Either a space inversion or a charge conjugation would reverse the correspondence chosen between helicity and sign of charge. However, the correlation between helicity and charge is invariant under combined space inversion and charge conjugation, the so-called CP transformation. Thus we see that abandoning space inversion in this way does not attribute to space any peculiar anisotropy or preference between left and right; rather space remains isotropic and a correlation between helicity and charge is introduced.

The requirements that neutrinos of only one helicity are emitted in a given type of decay requires that the neutrino have zero mass. If the neutrino had a finite mass its helicity could be reversed by going to a moving co-ordinate system with a high enough velocity to reverse the direction of the neutrino velocity. The helicity of the neutrino and therefore its lepton charge would depend upon the velocity of the observer and would be undefined in the rest system of the neutrino.

The two approaches to a two component theory are thus:

1. The Majorana neutrino, requiring zero charge and violation of lepton conservation. It can be consistent with space inversion and charge conjugation invariance and a finite neutrino mass.

2. The Lee-Yang, Salam or Landau neutrino, requiring zero mass and violation of space inversion and charge conjugation invariance. It can be consistent with lepton conservation and a finite neutrino charge.

6.6. SENSITIVITY AND LIMITATIONS OF EXPERIMENTS

In Chapter 3 we have seen how a complete set of experiments can be chosen to determine the lepton amplitudes for all the channels of allowed beta decay. In practice, any experiment always has a certain error. It is therefore of interest to go a bit further in our analysis of the different experiments to see how the accuracy of determination of any lepton amplitude is affected by experimental errors. In particular, since experiments indicate that most of the channels have zero amplitude, it is of special interest to see how sure we are that they are *really* zero and not just smaller than the other amplitudes.

The measurements of the angular momenta of the nuclear states are essentially exact. The nuclei are in eigenstates of the angular momentum and a measurement is either right or wrong but cannot be 'inaccurate'. The average helicities and the relative magnitudes and phases in mixed transitions are continuous variables. Each measurement specifies these quantities with a certain error, which is reflected in an uncertainty in the lepton amplitudes.

The simplest experiment is the measurement of the electron helicity or longitudinal polarization. Results for extremely relativistic electrons indicate a polarization of 100%, negative in negaton decay, positive in positon decay. The amplitudes for those channels having opposite polarization may be zero; however, they can also have a small finite value without bringing the polarization outside the limits of experimental error. In fact, since

$$\langle h_e \rangle = \frac{|a(h_e = +1)|^2 - |a(h_e = -1)|^2}{|a(h_e = +1)|^2 + |a(h_e = -1)|^2}$$

any deviation from 100% polarization would be only equal to twice the square of the amplitude for the channel of the opposite helicity. Thus the electron polarization measurement is not a sensitive test of possible small amplitudes of the other sign. A precision measurement of the polarization to 1%, giving a result as 100% ± 1% would still only place an upper limit of 7% on the amplitude for the channel with opposite polarization. Angular distribution measurements in pure Gamow-Teller transitions also measure the average helicity. They are therefore just as insensitive to changes in the amplitudes as the direct helicity experiment. They are also much more difficult in practice, because of the difficulty of producing polarized nuclei and measuring the degree of polarization, or of measuring the polarization of gamma rays.

One possible way to tie down the electron helicity more precisely would be to look for an interference effect which would be proportional to the amplitude, rather than the square of the amplitude. A measurement of transverse polarization of the beta rays would show such an interference effect. The transverse polarization would be zero if the longitudinal polarization were 100%. If the transverse polarization were measured to be zero, the limits of experimental error would place an upper limit on the amplitude of the lepton channel which is *coherent* with the dominant channel, and which has opposite electron helicity.

In order to obtain coherence it is necessary to choose an experiment in which only two channels can contribute, one dominant channel, and one having the opposite electron helicity but all other electron and neutrino quantum numbers the same as those of the dominant channel. A single measurement of the average transverse polarization of electrons emitted in beta decay should give a null result, as both values of m_e and m_ν contribute and exactly cancel out any interference effects. This is to be expected, as there is no preferred direction normal to the electron momentum if only the electron is measured.

The measurement of transverse polarization of electrons emitted at some definite angle with respect to the direction of polarization of a nucleus or gamma ray selects particular m_e channel and defines a preferred direction normal to the electron momentum. Such an experiment is difficult for the same reason as the angular distribution measurement discussed above with the additional difficulty of requiring an electron polarization measurement. However, if new experimental techniques for polarizing nuclei or detecting polarized gamma rays are developed, such an experiment may prove to be the best for placing upper limits on the small amplitudes, since the effect is proportional to the amplitude, rather than its square.

At electron velocities appreciably less than c, the mean electron helicity $\langle h_e \rangle$ is found to be in agreement with the value

$+v/c$ in $\beta^{\pm}$ decay predicted by beta decay theory for allowed transitions. Since the energy dependence of the mean helicity is predicted uniquely by the theory, any deviation from the value v/c at velocities less than c is directly related to a deviation from 100% polarization at $v = c$. The relation is just that given by (6.8), which includes the effects of Fierz interference. Since this interference is proportional to the *amplitude* of the channel which seems to be absent, rather than to the *square* of the amplitude, such a measurement gives a sensitive check on the presence of small finite amplitudes. The measurement of the longitudinal polarization at velocities appreciably less than c is therefore equivalent to the measurement of *transverse* polarization at $v = c$ discussed above in its sensitivity to small amplitudes, and is probably considerably easier. It does, however, require beta decay theory for its interpretation in terms of the amplitudes at $v_e = c$.

Since the Fierz interference terms appearing in the electron helicity are the same as those appearing in other Fierz interference effects, any experiment which determines the Fierz parameter b, (6.10), is equivalent to the electron polarization experiments in placing limits on the small amplitudes. Precise limits can be obtained from measurements of the ratio of positon emission to electron capture in decays where both processes are possible. The theoretical treatment of this case is beyond the scope of our book, but the results are of interest. These (SHERR [1954, 1956]; KONOPINSKI [1958]) place upper limits on the value of b of 2% for Gamow-Teller transitions and 10% for Fermi transitions. Such limits are better than would be obtained with a precision polarization experiment.

However, all the interference experiments discussed above can give information only about channels which can be coherent with the dominant channel; i.e. about channels in which the neutrino has the *same* helicity as that of the dominant channel. No interference experiment is feasible in practice between neutrino states of opposite helicity.

The measurement of the neutrino helicity or of the relative

helicity suffers from all of the difficulties mentioned above for the electron helicity, with the additional difficulties of indirect observation using measurements on a recoil nucleus and the practical impossibility of interference experiments. Except for the one case mentioned in Chapter 2, § 5 where the neutrino helicity was determined by direct measurement of the directions of the momentum and angular momentum of the recoil nucleus, the neutrino helicity has been determined only by electron-neutrino angular correlation experiments which determine the relative helicity. All these measurements involve the recoil nucleus and are extremely difficult. They measure a mean helicity or relative helicity and therefore any deviation from 100% is proportional to the *square* of the relevant amplitude. The sign of the helicity of the neutrino has been well established by these difficult and beautiful experiments. There seems to be little hope at present, for a precision experiment which will place an upper limit of even a few percent on the amplitude for opposite neutrino helicity.

The mixing ratio for the Fermi and Gamow-Teller amplitudes and their relative phase can be determined reasonably well by the lepton '$1 + A \cos \theta$' experiments. As has been discussed in Chapter 5, § 2 and 5, a precision determination of a phase at 0° or 180° requires more complicated triple correlation experiment which is sensitive to $\sin \phi$ rather than to $\cos \phi$. For the neutron decay, where important information about beta decay is obtained from these experiments, good results are obtained, and the triple correlation experiment tests time reversal invariance. In decays of complex nuclei these experiments give information about nuclear structure.

6.7. CONCLUSIONS

The status of the various beta decay experiments and the conclusions derived from them can now be summarized.

If the 100% polarization of the neutrino is accepted, either on theoretical grounds or by hopeful extrapolation of experiments, then the 100% polarization of electrons at $v = c$ follows to a very good accuracy from the absence of Fierz interference to the extent observed. Expressed in terms of the interactions of beta decay theory, the absence of Fierz interference leads to the conclusions that the Fermi interaction is *either* S *alone or* V *alone*, but not a mixture of the two, and that the Gamow-Teller interaction is either T *alone or* A *alone, but not a mixture of the two.*

It is interesting to note that these conclusions regarding beta decay interactions obtained from the absence of Fierz interference *are the same* as those obtained in the early days before the discovery of parity nonconservation. Although they are no longer valid in the general case where parity is not conserved, they become valid again when the 100% neutrino polarization is assumed. The reason for this is quite simple. Fierz interference occurs between two interactions which have opposite relative helicity in the extreme relativistic limit and the same values for all other quantum numbers. As long as there are two possible states for the neutrino helicity, the possibility exists for non-vanishing amplitudes in two channels having opposite relative helicity in the extreme relativistic limit and having different neutrino helicities. These states cannot give Fierz interference because of the difference in the neutrino helicity. The absence of Fierz interference therefore does not exclude the possibility that both these channels can have finite amplitudes. As soon as the number of neutrino helicity states is restricted to only one, whether by requiring 100% polarization, or by requiring a definite linear combination of helicity states having a definite *parity*, then the two interactions having opposite relative helicity in the extreme relativistic limit become coherent at lower velocities. If both are present, Fierz interference *must* be observed. The absence of Fierz interference then requires that one of the two interactions be absent.

Once both the electrons and neutrinos are believed to be fully polarized in the extreme relativistic limit, the *sign* of the electron polarization is determined by direct measurement and the sign of the relative helicity by the electron-neutrino angular correlation and by the one measurement of the neutrino helicity. The precisions of these measurements, are, however much lower than that of the conclusions from Fierz interference.

If the 100% neutrino polarization is not accepted, then the best way to put limits on the possible small amplitudes is to determine the maximum deviation of the electron polarization from full polarization by the direct experiments, and to use these in conjunction with the Fierz interference results to place limits in the deviation of the neutrino polarization from 100%.

Although the 100% neutrino polarization is far from well established with high precision, it is generally accepted. There does not seem to be any great incentive for experimentalists to improve the precision of difficult experiments in order to push the limits of error down a bit further. The two component theory of the neutrino with lepton conservation is very elegant. It does not seem to be reasonable that this theory should be valid only within the range of present experimental error, and that the other two states of the neutrino should exist and be present in beta decay to a much weaker extent. (This is of course a highly dangerous kind of argument!)

Further evidence for the opposite helicity for the leptons emitted in beta decay (i.e., for the (V, A) interaction) is obtained from the two decay modes of charged pions:

$$\pi^\pm \to \mu^\pm + \nu \tag{6.12a}$$

$$\pi^\pm \to e^\pm + \nu \, . \tag{6.12b}$$

If the interaction responsible for the emission of the leptons is assumed to be the same as that of beta decay, the leptons are emitted with opposite helicity in the extreme relativistic limit. At velocities less than c for the electron or muon, the amplitude for emission of the two leptons with the *same* helicity should be proportional to $\sqrt{\frac{1}{2}[1-(v/c)]}$, so that the mean relative helicity is $-v/c$. However, because the pion decay unlike ordinary beta decay is a *two-body decay*, the energy, momentum, relative helicity, and angular distribution of the two leptons are all determined uniquely by conservation of energy, momentum, and angular momentum, as shown in Fig. 6.5. In the rest system of the decaying pion, the

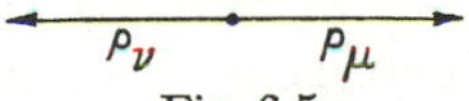

p_ν p_μ

Fig. 6.5.

two leptons are emitted in opposite directions with equal momenta, the magnitude of which is determined by energy conservation. Since the spin of the pion is zero, the angular momenta of the two leptons must be equal and opposite. Since both momenta and angular momenta are opposite, they must have the *same* helicity. Thus the usually dominant channel having opposite relative helicity is *forbidden* by the kinematics.

The total decay probability is then proportional to the square of the amplitude for the emission of the two leptons with the same helicity; i.e. to $\frac{1}{2}[1-(v/c)]$. Expressing this factor in terms of the mass m_π of the pion and the mass m of the charged lepton emitted in the decay, and using energy and momentum conservation, this factor becomes $1\{1+(m_\pi/m)^2\}$. If m is the muon mass, this factor is of the order of unity, but if m is the electron mass, this factor is very small, of the order of 10^{-4}. The probability of the muon decay mode is therefore very much higher than the electron decay mode, and this is born out by experiment. If a small amplitude were present for an interaction which gave the *same* relative helicity for the two leptons in the extreme relativistic limit, it would not be reduced by the factor 10^{-4} due to kinematic effects and should be easily seen in the π–e decay. The experimental results indicate that there is no such contribution and that the beta interaction consists exclusively of those types which give negative relative helicity in the relativistic limit. These results support the conclusions from the absence of Fierz interference in beta decay and the electron-neutrino angular correlation.

The relative magnitude and phase of the Fermi and Gamow-Teller (V and A) coupling constants are given by the '$1+A\cos\theta$' experiments on the decay of polarized neutrons. The relative phase is negative, as the electron asymmetry is small and the neutrino asymmetry is large. Indications that the phase shift is indeed 180° are obtained from the triple correlation experiment which places an upper limit on the direction from 180°. Since time reversal invariance requires that the phase be exactly 0° or 180°, it is reasonable to assume that beta decay is invariant under time reversal. *This can never be proved exactly*. Upper limits on the possible deviation from time reversal invariance can be decreased by improving the experimental accuracy.

The relative magnitude of the two interactions is then determined from the electron asymmetry. Since this asymmetry vanishes if $a_F = a_{GT}$, it is a sensitive measurement of the difference between them, which is about 20%.

REFERENCES

1. GELL-MANN, M., 1958, Phys. Rev. **111** 362.
2. GOLDHABER, M., L. GRODZINS and A. W. SUNYAR, 1958, Phys. Rev. **109** 1015.
3. KONOPINSKI, E. J., 1958, Theory of the classical beta decay measurements, Proceedings of the Rehovoth Conference (North-Holland Pub. Co.).
4. KONOPINSKI, E. J., 1959, Ann. Rev. of Nuclear Science, **9** 99.
5. LANDAU, L., 1957, Nucl. Phys. **3** 127.
6. LEE, T. D. and C. N. YANG, 1957, Phys. Rev. **106** 1671.
7. MAJORANA, E., 1937, Nuovo Cimento **14** 171, 322.
8. PAULI, W., 1957, Nuovo Cimento **6** 204.
9. SALAM, A., 1957, Nuovo Cimento **5** 299.
10. SCHIFF, L. I., 1955, Quantum Mechanics (McGraw-Hill Pub. Co.) p. 145.
11. SCHOPPER, H., 1960, Fortschritte der Physik **8** 317.
12. SHERR, R. and G. B. GERHART, 1956, Bull. Am. Phys. Soc. **1** 219.
13. SHERR, R. and R. H. MILLER, 1954, Phys. Rev. **93** 1076.
14. WU, C. S., E. AMBLER, R. W. HAYWARD, D. D. HOPPES and R. P. HUDSON, 1957, Phys. Rev. **105** 1413.

INDEX

Physics

OPTICAL RESONANCE AND TWO-LEVEL ATOMS, L. Allen and J. H. Eberly. Clear, comprehensive introduction to basic principles behind all quantum optical resonance phenomena. 53 illustrations. Preface. Index. 256pp. 5⅜ x 8½. 65533-4

QUANTUM THEORY, David Bohm. This advanced undergraduate-level text presents the quantum theory in terms of qualitative and imaginative concepts, followed by specific applications worked out in mathematical detail. Preface. Index. 655pp. 5⅜ x 8½. 65969-0

ATOMIC PHYSICS: 8th edition, Max Born. Nobel laureate's lucid treatment of kinetic theory of gases, elementary particles, nuclear atom, wave-corpuscles, atomic structure and spectral lines, much more. Over 40 appendices, bibliography. 495pp. 5⅜ x 8½. 65984-4

A SOPHISTICATE'S PRIMER OF RELATIVITY, P. W. Bridgman. Geared toward readers already acquainted with special relativity, this book transcends the view of theory as a working tool to answer natural questions: What is a frame of reference? What is a "law of nature"? What is the role of the "observer"? Extensive treatment, written in terms accessible to those without a scientific background. 1983 ed. xlviii+172pp. 5⅜ x 8½. 42549-5

AN INTRODUCTION TO HAMILTONIAN OPTICS, H. A. Buchdahl. Detailed account of the Hamiltonian treatment of aberration theory in geometrical optics. Many classes of optical systems defined in terms of the symmetries they possess. Problems with detailed solutions. 1970 edition. xv+360pp. 5⅜ x 8½. 67597-1

PRIMER OF QUANTUM MECHANICS, Marvin Chester. Introductory text examines the classical quantum bead on a track: its state and representations; operator eigenvalues; harmonic oscillator and bound bead in a symmetric force field; and bead in a spherical shell. Other topics include spin, matrices, and the structure of quantum mechanics; the simplest atom; indistinguishable particles; and stationary-state perturbation theory. 1992 ed. xiv+314pp. 6⅛ x 9¼. 42878-8

LECTURES ON QUANTUM MECHANICS, Paul A. M. Dirac. Four concise, brilliant lectures on mathematical methods in quantum mechanics from Nobel Prize–winning quantum pioneer build on idea of visualizing quantum theory through the use of classical mechanics. 96pp. 5⅜ x 8½. 41713-1

THIRTY YEARS THAT SHOOK PHYSICS: The Story of Quantum Theory, George Gamow. Lucid, accessible introduction to influential theory of energy and matter. Careful explanations of Dirac's anti-particles, Bohr's model of the atom, much more. 12 plates. Numerous drawings. 240pp. 5⅜ x 8½. 24895-X

ELECTRONIC STRUCTURE AND THE PROPERTIES OF SOLIDS: The Physics of the Chemical Bond, Walter A. Harrison. Innovative text offers basic understanding of the electronic structure of covalent and ionic solids, simple metals, transition metals and their compounds. Problems. 1980 edition. 582pp. 6⅛ x 9¼. 66021-4